Jan Volín, Pavel Šturm (Eds.)

HSCR 2021

Proceedings of the Fourth International Workshop on the
History of Speech Communication Research

Prague, August 27-28, 2021

TUDpress

Studientexte zur Sprachkommunikation
Hg. von Rüdiger Hoffmann
ISSN 0940-6832
Bd. 101

Jan Volín, Pavel Šturm (Eds.)

HSCR 2021
Proceedings of the Fourth International Workshop on the History of Speech Communication Research
Prague, August 27-28, 2021

TUDpress
2021

Cover image: Prague Castle through an arch of the Faculty of Arts
Copyright: Martin Hundák, Archive of the Faculty of Arts, Charles University

Bibliografische Information der Deutschen Nationalbibliothek
Die Deutsche Nationalbibliothek verzeichnet diese Publikation in der Deutschen Nationalbibliografie; detaillierte bibliografische Daten sind im Internet über http://dnb.d-nb.de abrufbar.

Bibliographic information published by the Deutsche Nationalbibliothek
The Deutsche Nationalbibliothek lists this publication in the Deutsche Nationalbibliografie; detailed bibliographic data are available in the Internet at http://dnb.d-nb.de.

ISBN 978-3-95908-434-5

Verlag der Wissenschaften
TUDpress ist ein Imprint von
THELEM Universitätsverlag und Buchhandlung GmbH & Co.KG
http://www.tudpress.de

Gesetzt von den Herausgebern.
Printed in Germany.

Preface

The series of international workshops on the "History of Speech Communication Research" (HSCR) was initiated in Dresden, in 2015. Its founders expressed their hope in published proceedings that the first event would be followed by more workshops of this kind in the years to come [1: 5]. Indeed, the *ISCA Special Interest Group: History of Speech Communication Sciences* supported the organization of the 2nd international workshop on HSCR in Helsinki, Finland, and the 3rd international workshop on HSCR in Vienna, Austria (see [2], [3]). The honour to host the current 4th event was bestowed to the Prague Institute of Phonetics.

The organizing institution is one of the oldest in Europe. It dates back to 1919, when the newly established state, Czechoslovakia, redesigned its university education and the Charles University officials of that period recognized the importance of the systematic approach to the phonic substance of speech. The first Czech professor of phonetics, Josef Chlumský, was given appropriate support and started the tradition which continues until today (see [4], [5]).

Prague is also known as the cradle of *functional structuralism*, i.e., a linguistic approach that emphasized communicative functions of language phenomena over the formalistic ones. The Prague Linguistic Circle was founded in 1926 and became a centre of progressive approaches to the matters of human language and its use in speech and in writing. The first generation of Prague linguists is still held in high esteem, even though their work was, due to the unfortunate historical development, not allowed to fully unfold. Both the Nazi and Communist regimes considered independent scientific activities dangerous to their grip on power.

The Prague Linguistic Circle renewed its activities after the fall of Communism. The current President of the Circle will deliver one of the keynote speeches at the workshop. In his speech, Tomáš Hoskovec is truthful to the Prague functional structuralist heritage, while at the same time emphasizing the perpetual concern of speech communication research: the link between the form and function, and the balanced examination of this link.

A tribute to the brilliant minds of the past is also central to the next three contributions to the agenda. Michael Ashby presents his research into the intellectual background of Frieda Goldman-Eisler. A brief but thrilling journey exposes generally unknown facts that put in doubt the presumable strong influence of Prague School, and especially of Karl Bühler on her approach to language research. In her paper, Markéta Vlková discusses the not fully appreciated legacy of Leoš Janáček's work on speech. The musical genius of the composer is appreciated by many, and his attraction to natural speech melodies is frequently mentioned by musicologists. However, Janáček's large collection of speech samples still awaits systematic and rigorous attention of speech scientists. Finally, Mária Gósy presents her findings concerning inspired work of Hungarian scholars and thinkers in the area of phonetic transcription. Again, we learn that history is much richer than it seems from cursory remarks in introductory chapters of common textbooks.

The second day of the 4th workshop, as well as the second part of these proceedings, are opened with the keynote speech by Angelika Braun. It is dedicated to several important real-life events – often quite dramatic ones – and their impact on developments in speech communication research. In a captivating way, the reader is led to appreciate the bond between society and science illustrated with specific historical events. The social context is also indicated in the paper by Jan Volín and Pavel Šturm, who use the case of intonational downtrends to demonstrate that historical achievements are often underestimated, and many authors refer incorrectly to the beginnings of research in a certain area. An identical concern is voiced by Jürgen Trouvain in his investigation of the vocal tract enlargement concomitant with voicing in stops. He, too, shows that the phenomenon was known and analysed earlier than commonly thought.

An interesting journey of discovery through various internal and external circumstances of their laboratory formation process is offered by Rüdiger Hoffmann and Dieter Mehnert. They explore the beginnings of speech communication research at Dresden Technical University with special attention to multiple influences that allowed for the foundation of the department.

The paper by Quintino Lopes and George Brock-Nannestad is clearly a tribute to the exceptional talents of Armando de Lacerda, who developed the technique of chromography to surpass the limitations of kymography. His success was outstanding and made an impact on many laboratories around the world. Here again, the link between phonetic instruments and social aspects of research can be observed.

Despite the current serious situation of the world pandemics and the resulting restrictions on conferences that have been changing from severe to moderate and back again for more than a year, we, the organizers, decided not to cancel the event. We persisted in parallel to the organizers of Interspeech 2021, of which the 4th international workshop on the HSCR is a satellite event. Our efforts thus resulted in yet another expression of admiration for the past achievements in speech communication research. Let us express hope that the series of workshops will continue in the future and the interest in history will grow.

Jan Volín and Pavel Šturm

Prague, July 2021

References

[1] HOFFMANN, R. & TROUVAIN, J. (Eds.): *HSCR 2015. Proceedings of the First International Workshop on the History of Speech Communication Research.* Dresden, September 4–5, 2015. Dresden: TUDpress, 2015.

[2] VAINIO. M., ŠIMKO, J. & AULANKO, R. (Eds.): *HSCR 2017. Proceedings of the Second International Workshop on the History of Speech Communication Research.* Helsinki, August 18–19, 2017. Dresden: TUDpress, 2017.

[3] PUCHER, M., TROUVAIN, J. & LOZO, C. (Eds.): *HSCR 2019. Proceedings of the Third International Workshop on the History of Speech Communication Research.* Vienna, September 13–14, 2019. Dresden: TUDpress, 2019.

[4] VOLÍN, J.: Speech sound structure studies in Prague: Differences in approaches and conflicts between methods. *La Linguistique*, 2014(2), pp. 83–100, 2014.

[5] ŠTURM, P.: The birth of an institute: A centennial jubilee of Prague's Institute of Phonetics. *Acta Universitatis Carolinae – Philologica*, 2019(2), pp. 9–26, 2019.

Contents

A HISTORY OF SOUND, SIGN, AND MEANING: THE PRAGUE CENTRE PERSPECTIVE AND PROSPECT

Tomáš Hoskovec

Cercle linguistique de Prague
cercle@cercledeprague.org

I.

In the unfortunate times of our present, in which even the most shameless lie may be called an alternative truth, acquiring thus legitimacy for subsequent public discussions, a public speech entitled *A history of* whatever – in this case, a history of sound, sign, and meaning – proves equivocal: will it be an alternative history in respect of what really happened in the past? Fortunately enough, that will not be the case. Nonetheless, we have to recognize that the history of whatever is always an intellectual construction: it is important that the construction be done honestly in an objectifiable way; it is important that the historical construct pay constant attention to the true nature of whose history is being constructed; and it is important that the historian never forget that history is *historically* a literary genre committed to illuminate our present understanding of our present times, which, historically again, has usually been conceived as illuminating men's own topical lives and acts through exempla carefully chosen from the past.

The historian of thought and ideas encounters the same crucial question as do his fellow-historians of states and countries, of social institutions and artistic movements, of cultures and languages, and, naturally, of particular human beings: what is the identity of their objects founded upon? In other words, what is the *same* in the constantly changing? The historian of thought and ideas searches in what people wanted to understand, and why; in how people understood their understanding, and what they held for the means to achieve it; in how they thought to recognize that they have already understood what they wanted to, and what attitudes they would adopt towards the results thus achieved. In respect of all that, the history of sound, sign, and meaning is a history of European scientific research into the nature of language communication.

It would be not only of small use, but utterly misleading to proceed by listing whatever has been meant, in the past, by the words *sound, sign, meaning*. On the contrary, any scientific history of a set of concepts (one isolated concept does not make sense, until opposed to some other concepts) requires a clear-cut a priori understanding, as well as the boldness not to take into consideration events and ideas that are traditionally studied under their headings unless they meet the aprioristic conditions consciously chosen. In this case, due to very particular philosophical circumstances, linguistics as a fully autonomous science could have never existed before the European 19th century started, and it may still totally disappear in this third decade of the 21st century, without even being noticed by current academia: there already are numerous chairs and departments of linguistics with no linguistics inside. All of that are consequences of the malaise which the institutionalized linguistics still prevailingly experiences when affronting – to say nothing about manipulating – the key-concept of the autonomous linguistics, the Saussurean linguistic sign, a unit having both sound and meaning.

The Prague Centre perspective and prospect, as the subtitle of the lecture announces, are the one of *functional structuralism* that the Prague Linguistic Circle has been developing for already one hundred years (in this consists the perspective), and the one of *encompassing philology* that largely draws on the achievements of the Paris School of interpretive semantics (in this consists the prospect). Encompassing philology – in Czech *celostní filologie*, German *Ganzheitsphilologie*, French *philologie englobante*, Italian *filologia totale*, Lithuanian *visuminė filologija*, Russian целостная филология – supplies adequate scientific instruments for grasping the evident and yet highly delicate empirical phenomenon of mutual human understanding in language communication (this constitutes its linguistic core); its instruments can then be adapted so as to make them effective for an inquiry into other than linguistic communication (this constitutes its semiological core). Since linguistic formations are being treated not only by linguistics and literary disciplines but also by numerous other humanities (in fact, virtually all, since all of them to a degree interpret certain written or oral texts as their sources), and since all humanities deal with semiological events (i.e., each of them interprets something that made a certain sense to a certain collective), encompassing philology is an auxiliary discipline for all humanities.

Encompassing philology, far from being some new-fangled invention of the contemporary Prague Linguistic Circle, is defined by a conceptual framework and a notional apparatus constructed by means of a conscious selection and purposeful composition of elements which have already come up in the history of linguistics at various points in time: the oldest ones about two hundred years ago (which, in fact, is not that long), the most recent within the last three decades (thanks to a critical reassessment of previous achievements); the original contribution of the proponent consists exclusively in their present rearrangement, which renders them newly topical.

II.

Encompassing philology draws on resources which have always been clearly present in the history of scientific study of language, notwithstanding the paradigms which may have at any single moment dominated institutionalized research. They can be listed as follows.

(i) The initial assumption needed for an emancipation of linguistics and semiology is a philosophical one: it was necessary to discard the Aristotelian-Scholastic prejudice that preformed thoughts come first, and only subsequently they are coded by various languages. The philosophical resolution was achieved by Immanuel Kant (1724–1804) and his authoritative account that a thought is the resultant vector of a process occurring entirely in the human mind; at the same time, he found an elegant retreat for the Aristotelian-Scholastic metaphysics by assigning it with a positive task: to study the inherent qualities of thought-processes.

(ii) However, an autonomous linguistics (and semiology) arises only once we realize that thinking itself is not only processual but also linguistic. This turn is performed by Wilhelm von Humboldt (1767–1835), historically the first to see and fully comprehend that a thought is essentially linguistic: it arises in language and by language; and since it is linguistic, it is, necessarily, also communicative; the birth of one person's thought in the milieu of a language is complete only once this same language calls up a response in another person; language is a social, which is to say, a cultural-historical phenomenon. Besides that, Humboldt is the first to see that linguistic units have to be understood as what we now call the Saussurean sign, i.e., as formations possessing expression and content *ex ante*, not created by a subsequent joining of an independent expression with an independent content; an expression is what it is by virtue of being the expression of a content, a content is what it is by virtue of being the content of an expression.

(iii) Step two is represented by Ferdinand de Saussure (1857–1913). Saussure confirms Humboldt's stance both regarding the relationship between thought and language and regarding the nature of the linguistic unit, which, for the first time, he calls *the sign*, and as a novelty – without actually ever using the term – he offers "structuralism". The essence of Saussurean structuralism is the mutual linkage of value and system: the value of any and all units is to be considered only as the resultant vector of the systemic relationship within which we observe the unit; and again, each system by consequence of the relationships that it contains provides the units involved in it with certain values. Saussurean structuralism is a necessary precondition for any scientific study of the Saussurean sign: rather than to refer, a sign attains values. Saussurean structuralism is not limited to linguistic signs; rather, from the very beginning, it also counts with sign systems of other kinds.

(iv) Step three is represented by Karl Bühler (1879–1963), a medicinal psychiatrist and experimental psychologist. Bühler (1927) neatly resolved the contemporary crisis of psychology by providing a positive programme, one which defines psychology not as study of the contents of human awareness or of psychic processes, but rather of the *meaningful living*: human beings, but not only humans, actively perceive the world they live in (*Umwelt*), accord meaning to what they perceive, and in accordance with the meaning they direct their behaviour in their world. In the spirit of this programme, Bühler (1934) proceeds to analyze the niveaus and links of significance in language communication in the way they are meaningful for humans (cf. Bühlerian functions, Bühlerian fields). Thus, the relationship between language and thought is explicitly transposed into an empirically verifiable relationship of communication and action.

(v) In its classical period (1926–1948), the Prague Linguistic Circle formulated and developed a programme of functional structuralism which integrates the insights (ii) through (iv): the structuralism of the Circle builds upon the Saussurean linkage of system and value, its functionality consists in accounting for the communicative aspect (in the Humboldtian as well as the Bühlerian sense). On top of that, the Circle supplies two highly practical solutions: first, it develops a concept wherein the sign is the entire oral or written text, while all the systemic elements – from phones all the way to thematic patterns – serve as diacritical moments, structurally differentiating one large sign from another; second, it develops a concept whereby the extralinguistic reality is always being confronted not by an isolated word but rather by the entire text, i.e., a sign evincing extreme structural complexity and highly structured meaning; thus, the barrier of ontological reference is eliminated.

(vi) The Paris school of interpretive semantics attained an elegant solution of the inherent meaning dynamics of small sign units, especially lexical ones, in their textual occurrences. If the meaning of each sign unit is determined by a defining domain, i.e., a systemic section wherein it is being viewed, then the concretization of meaning in the entirety of a text consists in the actualization of one defining domain out of many potential ones, and this choice is being directed by the global structure of the text. The technique of interpretive semantics, as developed by François Rastier via a critical reassessment of the broader French tradition (Bernard Pottier, Algirdas-Julien Greimas), evinces deep parallels with the older techniques of functional structuralism, as they were developed primarily by Jan Mukařovský and Jiří Veltruský. For both the Paris and the Prague stance it is symptomatic that they combine the two concepts of language which have coexisted since Humboldt and Saussure, i.e., language as an abstract system, and language as a social institution; social institutions guide the actualization of defining domains, which are, in their potentiality, deposed in an abstract system.

(vii) Encompassing philology intelligently evaluates all the historical experiences listed above and transforms them into practical instruments, which allow for both describing a language and for analyzing a linguistic formation, i.e., a text. Unlike French semantic tradition (from Benveniste to Rastier) which determines, by mutual differentiations, the value of small linguistic signs only, mostly morphemes and lexemes, leaving the rest to what Émile Benveniste calls *mode sémantique*, encompassing philology disposes of systemic means for differentiating large signs as well, including entire texts. In principle, such means can be adopted even to non-linguistic signs, i.e., to cultural-historical events of various nature (cf. Jiří Veltruský, TCLP. n.s. 6). Encompassing philology was launched internationally by Tomáš Hoskovec; its most recent international platform is presented in the *Thèses de Prague 2016* (TCLP, n.s. 8), which have already been discussed, among others, in a trans-Atlantic tutorial Prague-Calgary-Bogotà-

Abidjan. It has been successfully applied to historical-cultural and historical-linguistic studies by Ilja Lemeškin (TCLP, n.s. 10).

III.

Under the circumstances described above, the lecturer seeks, first, to demonstrate that present-day humanities provide room for working with the language sign (in Saussure's sense) in a way semiologically correct, philologically exhaustive and heuristically profitable, and second, to point out that the crucial elements needed for such work have been available for quite some time; the emphasis is put on demonstrating practicability in execution. This presentation proceeds in three steps.

First of all, we must eliminate the principal barrier for a historical understanding of European structuralism (there was a U.S. structuralism as well) and its relationship to semiology. This barrier has been erected not by lack of knowledge, but rather by the preponderance of superficial, or indeed false knowledge. European structuralism has to be understood as *corpora* of scientific texts, produced during more than a century of varyingly intense and variously interrupted activity. These corpora are not "just there"; there are always consciously compiled by a researcher responsible for his selection. The criterion of the present selection is, whether and how scientific research comes to terms with the principles of Saussurean structuralism and Saussurean sign, as described above under II. (ii)–(iii). Within the scientific production thus selected, we discern *centres*, i.e., large collections of texts that are aware of one another and recognize certain shared goals, even though their procedures may differ. Methodological unity of the conceptual and interpretive apparatus is the defining quality of a yet narrower collection of texts, called *school*; as a rule, there are several schools in each centre (and the "Prague School" label as traditionally employed actually designated the Prague Centre of functional structuralism, within which numerous schools may be discerned). The lecturer will characterize the thematic specificity of the important centres of European structuralism (Prague, Geneva, Copenhagen, Paris) and describe their programmatic development as well as the modifications of the dynamics of their research production.

In the following step, it is indispensable to exemplify a systemic description of a language as a sign system. The crucial quality is the assumption of a top-to-bottom perspective: large signs (such as, in particular, entire texts) are not composed out of smaller ones and smaller ones are not composed out of elementary ones; rather, the large and global sign is always primary, and only secondarily does it disintegrate into smaller signs, where the separation of lower-level, partial signs must always be legitimized by detecting their structural relationship to the higher-level whole. This is how variously voluminous textual passages – and ultimately single utterances qua elementary texts – separate out of a global text. Within the further structural descent, the analysis introduces the structurally-analytical unit "sentence", defined completely independently from "utterance", and only against the background of the structural unit "sentence" there come forward such sign units as "word" and "grammatical relationship", as well as diacritical units such as "prosodic contour" or "phoneme" at the expression plane, "seme" at the content plane.

Finally, we must demonstrate a model of grasping the meaning of the text in an objectivized way by employing a dynamic selection of semes. The content units of a text are modelled as configurations of semes, i.e., the minimal diacritical moments effective in the content plane of a language. The occurrence of a seme is demonstrable by the actualization of the pertinent defining domain as against other, systemically compossible defining domains. The choice of defining domains is being determined both by the structural relationships within the given text, and by the cultural-historical anchoring of linguistic signs in a particular collective.

IV.

To conclude, encompassing philology offers both practical assistance and improved scientific self-confidence to the Humanities (*Geisteswissenschaften, sciences humaines*), denigrated as they often are in a presumed contrast to the "hard", "exact", truly "scientific" disciplines, a contrast currently mirrored in the standard usage of languages such as English (*sciences* :: *humanities*) and French (*sciences* :: *lettres*). The essence of scientific exactness does consist not in making exact measurement but rather in separating certain structural components, including those to be measured, out of the empirical reality, and then interpreting them in a methodical manner with respect to the behaviour of the whole. In this sense, there is only one science; and the scientific community, being simply one out of many various collectives within humanity, is characterized by the fact that within it, the specific norms of rational thought and objectivized procedures are especially dominant.

References

As underlined in the speech, any history of whatever shall be founded in exhaustive corpora of texts. For the history of sound, sign, and meaning, the corpora are of such an amount that not even a sketch of them could be made here. The Prague Linguistic Circle has been elaborating, within a large-scaled international research project called *Atlas du structuralisme européen*, an open-source basis of structured and annotated bibliographies: all who may feel concerned are kindly invited to join these efforts. Instead of references, a list of the last issues of the new series of *Travaux du Cercle linguistique de Prague* is given, in which materializes another aspect of the *Atlas* project; an outline of the historiographic-bibliographic basis for the classical period of the Prague structuralism can be found in [9], another one, for the classical Prague phonology, in [14]. Above that, a list of proponent's selected papers is added, in which the programme of encompassing philology is being materialised.

Author's own texts explaining the past and prospects of functional structuralism:

[1] HOSKOVEC, T.: Synchronie et diachronie au centre et à la périphérie du foyer pragois de structuralisme fonctionnel. *Cahiers Ferdinand de Saussure*, 64, pp. 49–72. Genève: Droz, 2011.

[2] HOSKOVEC, T.: La linguistique textuelle et le programme de philologie englobante. *Verbum* 22(2), pp. 193–218, 2012. Presses universitaires de Nancy.

[3] HOSKOVEC, T.: Jiří Veltruský (1919 – 1994). A journey through life, with semiotics. In: *TCLP*, n.s. 6, pp. 206–225, 2012.

[4] HOSKOVEC, T.: La perspective fonctionnelle de la phrase dans une perspective historico-philologique du foyer pragois de structuralisme fonctionnel. *Écho des études romanes*, 8(1), pp. 7–25, 2012. České Budějovice: Jihočeská universita.

[5] HOSKOVEC, T.: Poetika a filologie [L'ars poétique et la philologie]. In: A. VEŘMIŘOVSKÝ (Ed.): *Poetický Cikháj v Brně 2010. Sborník textů o poetice.* Brno: Tribun EU, pp. 13–27, 2012.

[6] HOSKOVEC, T.: Strukturalismus a sémiotika [Structuralisme et sémiotique]. In: L. MARTINEK (Ed.): *Český a polský strukturalismus a poststrukturalismus – historie a současnost.* Opava: Slezská universita, pp. 121–141, 2015.

[7] HOSKOVEC, T.: Co chtít od sémiotiky? Zamyšlení nad skonem Umberta Eca [Que faut-il et que peut-on exiger de la sémiotique? Réflexion à l'occasion du décès d'Umberto Eco]. *Romboid*, 51(9), pp. 47–52, 2016. Bratislava.

[8] HOSKOVEC, T.: Faire entrevoir l'Europe aux Européens à travers leurs langues. Réflexions en guise de compte-rendu du livre « George J. (Jiří) Marvan: Introducing Europe to Europeans through their language ». In: P. ZIMA (Ed.): *Europe and the Others*. München: Lincom, pp. 135–138, 2016.

[9] Hoskovec, T.: Karl Bühler et le programme sémiologique du Cercle linguistique de Prague. In: *TCLP*, n.s. 7, pp. 335–385, 2018.

[10] Hoskovec, T.: Письменность Великого княжества Литовского как европейский феномен. О культурно-историческом контексте восприятия Франциска Скорины [La production littéraire du Grand-Duché de Lituanie comme phénomène européen. Le contexte historique et culturel de la traduction de la Bible en langue ruthène par le docteur Francisk Skorina (1517)]. *Lietuvos mokslų akademijos Vrublevskių biblioteka 2013 – 2014 (Metraštis)*, pp. 144–153, 2018. Vilnius.

[11] Hoskovec, T.: Jazyk, ideální slovníkové heslo [Langue et langage, entrée encyclopédique, version idéale]. *Litikon*, 3(2), 2018. Nitra.

[12] Hoskovec, T.: Préface. In: *TCLP*, n.s. 8, pp. 9–18, 2019.

[13] Hoskovec, T.: Thèses de Prague 2016. In: *TCLP*, n.s. 8, pp. 21–38, 2019. Previously published in *Texto! (Textes & Cultures)*, 22(1), 2018, http://www.revue-texto.net/index.php?id=3842.

[14] Hoskovec, T.: Sentence prosody and experimental phonetics. The way out of the *cul-de-sac* of phonology. In: *TCLP*, n.s. 8, pp. 101–122, 2019.

[15] Hoskovec, T.: … und wenn die Sprache einmal gebührend abgegrenzt wird? *Linguistische Treffen in Wrocław*, 15(1), pp. 101–114, 2019. Previously published in *Texto! (Textes et cultures)*, 23(2), 2018, http://www.revue-texto.net/index.php?id=4047.

[16] Hoskovec, T.: Préface. In: *TCLP*, n.s. 10, pp. 7–18, 2020.

[17] Hoskovec, T.: Lithvanica aliter. Visuminės filologijos pritaikymas lituanistikoje [Lithvanica aliter: encompassing philology as to be applied to Baltic studies]. *Acta Linguistica Lithuanica*, in press.

Travaux du Cercle linguistique de Prague, nouvelle série (TCLP, n.s.):

[18] **TCLP, n.s. 6.** = Jiří Veltruský: *An Approach to the Semiotics of Theatre*. Edited by D. Drozd & T. Hoskovec. Brno: Masaryk University, 2012.

[19] **TCLP, n.s. 7.** = *Karl Bühler, une théorie du langage redécouverte – eine Sprachtheorie wiederentdeckt – a theory of language rediscovered*. Edited by T. Hoskovec, S. Raynaud, F. Albano Leoni & J. Trabant. Praha: OPS–PLK, 2018.

[20] **TCLP, n.s. 8.** = *Expérience et avenir du structuralisme – Vergangenheit und Zukunft des Strukturalismus – The past and prospects of structuralism*. Edited by T. Hoskovec. Praha: OPS–PLK, 2019.

[21] **TCLP, n.s. 10.** = Lemeškin, I.: Портрет Франциска Скорины – *Pranciškaus Skorinos portretas – Portrait de Francisk Skorina*. Edited by T. Hoskovec. Vilnius–Praha: OPS–PLK, 2020.

[22] **TCLP, n.s. 9, in preparation.** Hedánek, J.: *An Outline of Phonology*.

A PRAGUE SCHOOL PSYCHOLINGUIST IN LONDON? THE LIFE AND CAREER OF FRIEDA GOLDMAN-EISLER (1907–1982)

Michael Ashby

Department of Speech, Hearing and Phonetic Science, UCL, London, UK
m.ashby@ucl.ac.uk

Abstract: This paper sketches the life and work of Frieda Goldman-Eisler, a pioneer of experimental psycholinguistics who worked in the Department of Phonetics at University College London and became the first Professor of Psycholinguistics in the UK. We trace the evolution of her work in the intellectual climate of the 1950s, and assess the likely influence of her colleagues at UCL. We challenge the widely-held assumption that she was influenced by the Prague-circle psychologist Karl Bühler. Drawing on previously unused biographical and archival sources, we give details of her little-known first marriage and hapless academic record as a postgraduate student in London.

1 Introduction

Frieda Goldman-Eisler (1907–1982) published an extensive series of papers in the 1950s and 1960s in which she utilized experimental studies of the temporal structure of spontaneous speech—particularly the location and durations of pauses—as quantitative evidence in the investigation of the internal processes underlying its production. She wrote that 'Psycholinguistics was neither name nor concept at the time and the few workers who busied themselves in this area were very much on their own and cut off from the main stream of psychological thought' [14: 5]. It is true that the term was a relatively new coinage, and there were as yet no laboratories or centres devoted to the field, or any established paradigms for research. To that extent she was a pioneer, and the Psycholinguistics Research Unit which was set up within the UCL Phonetics Department to provide a base for her and her students was for a time unique.

Her earliest publications, starting in 1948, are under the name Frieda Goldman. She was then in her early 40s and had a doctorate (Dr.Phil) gained in 1931 from the University of Vienna, where her teachers had included Karl Bühler, the psychologist and philosopher of language whose work influenced the Prague linguistic circle. In 1950, she married Paul Eisler (1907–1992), an engineer and inventor, who like her had fled the growing Nazi threat in Austria in the 1930s and made a career in the UK. Following the marriage, Frieda published under the name Goldman-Eisler.[1] Paul Eisler is chiefly notable as the inventor of printed circuit technology, which revolutionized electronics.

From 1955, and for the rest of her career, Frieda was associated with UCL. At first, her status was described as 'Honorary Research Assistant'. The next year she was styled 'Honorary Research Associate' and this 'honorary' (i.e. unpaid) status continued until 1961. During these years, research funding came from the Medical Research Council, and later from its US counterpart, the National Institutes of Health. It was only in 1962, when she was 55 years of age, that she joined the payroll as a Lecturer. Thereafter she was rapidly promoted to Reader (1966) and to Professor (1971). She retired in 1974, with the title of Emeritus Professor,[2] becoming an Honorary Research Fellow, retaining an office at UCL and for a time continuing her research.

Frieda was the first holder of an academic post in the UK explicitly linked with psycholinguistics, and certainly the UK's first Professor of Psycholinguistics. The mid 1960s were the high-point of her celebrity and influence, although she continued to research and publish for another ten years. After a long illness, she died from a brain tumour in 1982.

2 Early papers, 1951–1955

Over the period 1948–1955 the Medical Research Council funded a programme of work at the Maudsley Hospital in London (the UK's leading psychiatric training centre) which was designed to investigate factors influencing the effectiveness of psychotherapeutic interviews. Frieda was engaged for this research—which was not of her own design—almost certainly not because of any previous expertise with spontaneous speech but rather because her work up to that point similarly involved the empirical testing of (then very influential) psychoanalytic ideas.

The focus of the research she was carrying out was not on the unquantifiable content of *what* was said, but rather *how* it was said: 'the speed of talking, the pauses, silences, interruptions, the rate of talking', etc., and the paralinguistic accompaniments ('the expressive movements accompanying it, gestures, facial expressions, and mannerisms') As she pointed out, these occurrences and their durations were functions of time and potentially lent themselves to quantitative treatment [10: 355]. This approach closely mirrored that pioneered in the United States a few years earlier by Chapple [3].

Chapple's method of working entailed the live observation of an interview from behind a one-way screen, the observer being equipped with a special event-recording device of Chapple's own devising, which he termed the 'interaction chronograph'. In its simplest form it had a small number of keys resembling those on a typewriter, which when pressed made marks on paper being fed from a roll at constant speed. Magnetic tape recording was not yet widely available and the much less convenient method of disc recording was not routinely employed.

Even if the task were to register simply whether A or B (or neither) was talking at any given moment, the method relied on a skilled operator pressing the appropriate keys at the right times. Frieda specifically thanks 'last but not least Mrs M. Malloy for her invaluable assistance in this experiment and the skill with which she recorded the conversations' [10: 362]. Of course, producing the chronograph recordings was only the beginning. The subsequent analysis of the resulting paper roll records was laborious: a half-hour interview took about 15 hours to analyse. Further labour was then required for statistical processing of the numerical data and the plotting of graphs to display results.

3 An interdisciplinary home in the Phonetics Department, UCL

By 1955, if not earlier, Frieda had come into contact with the Department of Phonetics at University College London. We don't know how the connection was first made, but there are many possibilities. It may have come about via people she knew in the UCL Psychology Department. Or perhaps she sought out the Phonetics Department because her own work at this time took a turn in the direction of speech physiology. A paper of 1955 (actually completed in the summer of 1954) investigates breathing pattern as an indicator of emotional state in the speaker, and the study of respiration in speech was a long-established topic in phonetic research.

Following the retirement of Daniel Jones, the Phonetics Department was now under the direction of D. B. Fry (1907–1983). Frieda describes how in 1955 Fry 'offered her a home' in the Department [14: v]. This was not a salaried position, and in practical terms amounted at first to little more than use of some office space and a postal address.

But probably more important than the physical pied-à-terre was a congenial and stimulating collegiate atmosphere, and the opportunities for networking which it brought. This was undoubtedly encouraged and fostered by Fry himself, who championed an interdisciplinary approach to the study of speech and communication. As early as 1946 he had succeeded in appointing an electrical engineer, Peter B. Denes (1920–1996), to the otherwise arts-oriented staff of the Phonetics Department. During the 1950s, Fry's interdisciplinary outlook was to provide the first academic footholds in the UK for both psycholinguistics and sociolinguistics.

An informal focus for much of the interdisciplinary interaction at UCL in those years was provided by the senior staff common rooms of the College. In the 1950s and 1960s academics still had time to stop work for a cooked lunch in one of the college refectories, and then relax together for a while over coffee or drinks in the large armchairs or sofas, or around tables laid out with the day's newspapers and recent issues of leading cultural and scientific periodicals such as *Nature*. We know that Frieda made use of these facilities, and profited from the connections she made. In his autobiography, Paul Eisler describes how Frieda met mathematician and engineer E. H. Thompson (1910–1976), who was Professor of Photogrammetry at UCL, and introduced him to her husband. The two men struck up a strong rapport, and the three went into business, founding Eisler Consultants which operated from a laboratory established in the Eislers' large North West London home, to develop and market Eisler's patents [4: 91].

4 The intellectual climate

But there was more to interdisciplinarity than lunchtime chat. It was a serious and extensive academic fashion, and gave rise to academic exchange visits, conferences, and research centres. It was associated, too, with another of the big ideas of the day, communication theory. Starting in 1953, a group of senior academics from across the whole range of UCL's departments and faculties organized themselves into what became the Communication Research Centre. One of its early activities was a public lecture series which was published in book form [2]. To this Fry contributed a substantial review paper 'The experimental study of speech' [6]. In search of more technical expertise on the subject of information theory than could be found in UCL at the time, the group invited a contribution from Colin Cherry, who was at Imperial College, and he gave a paper with the title 'Communication Theory and human behaviour'. Perhaps Frieda attended some of these talks, and thus encountered Fry as a result. As a psychologist working in the field she was, she would almost certainly have known Cherry's then newly-published work on selective attention in listening—the so-called 'cocktail party effect'.

The Communication Research Centre cast its net widely. *Studies in communication* [2] contains not only fairly predictable chapters dealing with the philosophical and mathematical analysis of communication and with phonetics and language description, but more surprisingly 'Communication in Biology', 'Interpretation of Visual Symbols in the Arts', and even 'Communication in Economic Systems'. Indeed, the group might appropriately have styled themselves a Centre for Semiotics. A 1960 description in the college *Calendar* says that the purpose of the Centre is '… to investigate the usage and efficacy of signs and symbols—especially those of a linguistic kind'; and adds 'in this it takes account not only of relevant work done by students of language, but also of the contributions made by students of (for example) psychology, philosophy, physiology, engineering, and the visual arts' (page 428).

The mathematical theory of communication developed by Shannon played an important role in triggering a wide interest in communication, though the idea of a connection between mathematical foundations and the numerous projected applications was probably more of a pious hope than anything real. Very few could understand Shannon in any depth; he soon moves on to theorems and proofs which are well beyond the non-mathematician [20]. But the reassuring suggestion of a rigorous mathematical underpinning for what would otherwise be rather nebulous investigations into communication nevertheless had a particular appeal in what was an era of behaviourism and operationalism. Even the genial and expansive Fry paid lip-service to the prevailing orthodoxy: 'It will be clear that an experimental study of speech presupposes a behavioural view of the subject…' [6: 147].

5 Developments in Frieda's research at UCL

Chapple's methodology for the study of interviews can be seen as operationalism pushed to the extreme. All the complexity of an extended human interaction is emptied of meanings and reduced to a pattern of on-off timings of activity. It was this paradigm which provided the starting point for Frieda's research. Chapple's contention was that the temporal patterns which emerge are consistent, characteristic of individual participants, and indicative of aspects of their personalities or states of mind, and Frieda's early work was directed at exploring these claims.

But from 1955 onwards her work developed in several new directions, some of which were clearly influenced by her contact with Phonetics. In the 1955 paper on breathing, she monitored breathing non-invasively by listening for inhalation sounds with 'a sensitive microphone' [10: 54]. This seems to be the first mention of a microphone in her publications and probably indicates that by this date she was using magnetic tape recording in her work. High-quality tape recording was to become the norm, and she made many of her later experimental recordings in the ideal acoustic conditions of the Department's newly-built anechoic chamber, which had been completed in 1948.

5.1 The statistics of speech

Though the detail of Shannon's mathematical analysis of communication is beyond many readers, the attention even of a casual reader is likely to be caught by the section where Shannon introduces the concept of statistical structure in language [24: 43]. Eye-catching blocks of text demonstrate the progression from a zero-order approximation of English which begins 'XFOML RXKHRJFFJUJ ZLPWCFWKCYJ FFJEYVKCQSGHYD …' through a series of increasingly close approximations, to a text which is no more than a second-order word approximation but seems to teeter tantalizingly on the verge of making sense:

> THE HEAD AND IN FRONTAL ATTACK ON AN ENGLISH WRITER THAT THE CHARACTER OF THIS POINT IS THEREFORE ANOTHER METHOD FOR THE LETTERS THAT THE TIME OF WHOEVER TOLD THE PROBLEM FOR AN UNEXPECTED.

Shannon used the letter- and word-frequencies of printed English but the idea can be extended to phoneme frequencies.

A favourite demonstration in Fry's lectures, based on Shannon, was to play a guessing game with the audience. He would announce that he had in mind an English sentence, and the audience was to guess each of the phonemes it contained, starting with the first. As the guesses were called out, he kept a tally on the chalkboard of the number of guesses required in each position and drew a bar-chart, to show the fluctuating—but generally declining—number of guesses required as more and more of the sentence is revealed and those guessing can draw on more and more of their linguistic and contextual knowledge. A typical example is contained in Fry's lecture to the Communication Research Centre [2: 152].

We cannot be certain that it was Fry who introduced Frieda to a concern with a statistical structure in speech, but even if he were not the source he would certainly have encouraged the interest.

Her 1957 letter to *Nature* [12] announces Frieda's most strikingly original observation—that hesitation pauses in speech are located at points of low transitional probability (and hence at points of high information content). She estimated the word-transition probabilities by having further experimental subjects make guesses from written versions of the originally spoken sentences. Subjects for these experiments, as often, were Frieda's colleagues in the Department of Phonetics. This was a time when experimenters saw no need to go further than the office next door or into the corridor to find their subjects.

5.2 Cognitive processing in speech

The 1957 note, and the longer account of the same experiments [13], also mark a shift in interest from speech viewed solely as a diagnostic tool for psychotherapy towards a concern with the mental planning of speech and the process of word-finding. For the first time she references Hughlings Jackson (1885–1911) whose ideas on language were to influence her greatly.

The clinical connection was however always retained in some measure. In later work she collaborated with Andrew Skarbek (1925–2011), a physician and psychotherapist, on the effects of psychoactive drugs on speech production. Professor J. C. Wells (b. 1939) recalls serving as an experimental subject when he was first a postgraduate student in the Department (1960–1962). He was required to make recordings in the anechoic room having been injected by Skarbek with the antipsychotic drug chlorpromazine.

5.3 Automatic pause detection

The methodology which Frieda adopted from Chapple depended on skilled and labour-intensive human logging of a spoken interaction. Starting in 1958 Frieda began to use a specially designed device to discriminate pauses automatically from speaking activity. The device (essentially a rectifier followed by a trigger circuit) makes a categorical on-off choice, turning any detected speech into a standard wave of constant size. Later a two-channel version was produced, as required for her study of simultaneous translation.

Although such a device is in principle very simple and might suggest itself to any engineer to whom the problem was posed, it can hardly be an accident that it also resembles the very first stage of the speech recogniser then under development at UCL by Fry and Denes [7]. As they explain, the recogniser's first task is to distinguish speech from silence.

The output of Frieda's speech/pause detector was led to a pen recorder (see Figure 1). The paper record from this, typically on a long roll of grey 'Teledeltos' paper, had then to be manually analysed as before—a laborious and painstaking measuring task, delegated where possible to an assistant. In later work this phase too was partly automated. Frieda gained the use of a digital counter/timer and a digital data logger—both, at the time, uncommon and expensive items.

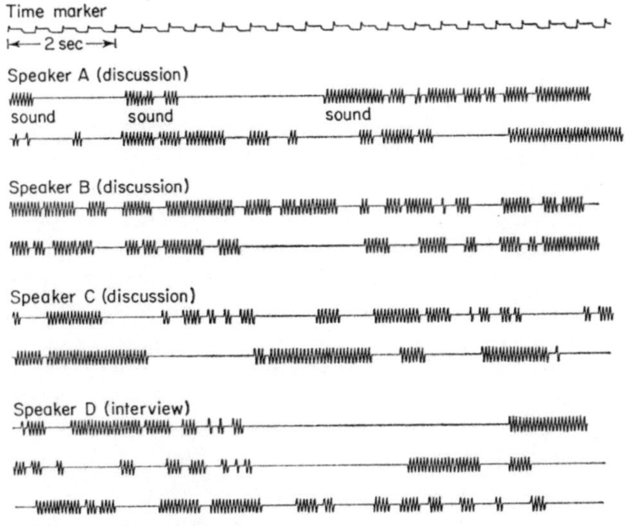

Figure 1. The first illustration from Frieda's 1968 book *Psycholinguistics* [14: 11], which is captioned 'Graphic transformations of spontaneous speech by four speakers'. The same figure was used on the dustjacket of the book.

A new version of the 'speech analyser' was built, operating essentially as before, but now producing accurately timed pulse trains which were sent to the counter. Each count was then digitally logged—but there the automation stopped. The output from the data logger was a long roll of paper resembling a supermarket till receipt, each line being a print-out of the duration of a single detected event. It was still a time-consuming task for somebody to work through the printed list and collate the results.

By the late 1960s Frieda was doing the statistical processing of her results on the university's huge mainframe computer, so another painstaking and laborious job for the research assistant was that of keying the punched cards from which the data was read. By 1970 both the Phonetics and Psychology departments had acquired their own first laboratory computers (respectively a DEC PDP-12 and a PDP-8), but there is no indication that either was applied to Frieda's research.

6 Frieda's output

As is remarked by O'Connell & Kowal [19], Frieda was very prolific. There seems to be no comprehensive bibliography of her output, but O'Connell & Kowal's list (which is definitely incomplete) runs to more than thirty publications. Over several years, Frieda must have been one of the most productive members of the department. It is hardly possible to summarise her work in the present short paper. Her book *Psycholinguistics: Experiments in spontaneous speech* [14] is essentially her own reprise of her work up to 1968. The majority of her papers have a narrowly-defined focus but when it was offered she clearly relished the opportunity to write more expansively and draw on wide reading, as in her invited contribution to the Lenneberg volume [16].

Many of her publications appeared in psychology journals. In the 1950s there were few journal outlets for experimental work concerned with speech. To help fill this gap Fry founded the journal *Language and Speech* in 1958 and Frieda joined the editorial board. Issue 1 of the new journal contains Frieda's 'Speech analysis and mental processes' and thereafter she became a frequent contributor. The same inaugural issue contains 'The solution of some fundamental problems in mechanical speech recognition' by Fry and Denes—the very paper which includes their discussion of the problem of automatically dividing speech from silence.

7 Frieda and Bühler

We turn now to the question of how much Frieda took from her training in Vienna, the extent to which she was influenced by Bühler, and the extent, therefore, to which she might count as a link between the Prague School and UCL.

As a student in Vienna, she would have attended Bühler's lectures at the time when he must already have been working on his major work *Sprachtheorie*, published in 1934. She completed her own Dr.Phil in 1931—the very year in which Bühler published a paper 'Phonetik und Phonologie' in the *Travaux du cercle linguistique de Prague*. Superficially, the coincidences and correspondences are striking, and many have understandably been tempted to assume that Frieda emerged from her Vienna training as a qualified psycholinguist from the outset. In a table categorizing emigrant Austrian women by specialism, Prost [21: 451] lists Frieda as a psycholinguist. Hinton [15: 110] even speculates that Frieda's linguistic training (i.e. presumably what she had learned in Vienna) might have fitted her for phonetic fieldwork in the Jewish East End which was briefly contemplated in 1939.

The first thing to note is that Frieda simply doesn't cite Bühler. The sole mention I have been able to find [14: 95] concerns pauses inserted consciously for rhetorical effect—that is, precisely what her research is *not* dealing with. And in a passage where she's acknowledging the psychologists and philosophers who most influenced her general outlook, she gives an alphabetical list that runs from Bergson to Wundt—but Bühler is not one of them [14: v].

While her course of study in Vienna did contain psychology, the main focus of her work was literature, and her approach to that was conditioned by her strongly-held socialist views. As library catalogues reveal, the 1931 Dr.Phil has as its subject the social satire of Nestroy.[3] In the conditions of the Covid-19 pandemic it has not been possible to see the thesis, but Maas [18] had evidently examined the typescript and had also seen some of the documentation relating to the examination. He reports that the thesis attempts a Marxist interpretation, which met with a lukewarm response from the faculty. In the oral examination, Bühler awarded only the mark *genügend* (i.e. 'sufficient').

8 Frieda's early life and first marriage[4]

Frieda's early life, and her activities in the 1930s and the war years, are also relevant to the question of whether there is any continuity between her studies in Vienna and the psycholinguistic experimentation for which she became famous in the 1950s.

She was born Frymet Leib in 1907 in the city of Tarnów which is in modern-day Poland but was then in Galicia, a territory within Austro-Hungary. She was the third and youngest child of Adolf Leib, a textile manufacturer in Vienna. Her mother Pepi came from Sambor (Sambir) which is in modern Ukraine. The family was German Jewish. She attended Vienna University over the years 1925–1931, gaining the degree of Dr.Phil. She was a member of the Socialist Students' Association and of the Social Democratic Party of Austria (and from various indications we may surmise that she was in fact a member of the Communist Party). Significantly, she worked as a film critic for the Vienna edition of the celebrated weekly *Die Weltbühne*, which mixed radical politics with the arts and economic affairs. Attempts to ban the publication and prosecute its editors had started in Germany as early as the 1920s.

Figure 2. Frieda as a young woman. Photographer and date unknown but probably c.1935–1937. She is dressed in the traditional *dirndl*, a fashionable choice for a special occasion or celebration in Vienna. Reproduced by kind permission of Emma Goldman.

She gained the status of British subject in 1934 when she married the writer Willy Goldman (1910–2009), then unknown but later the author of *East End My Cradle* (1940) and several other notable fictional works set in London's East End (for brief accounts of his life and work, see [9], [5]). The marriage was entered into at least partly to give her a means of escape from the Nazi threat, though there seems to have been genuine affection too, at least at first. They certainly had fervent cultural and political interests in common—and for the rest of his life, he would credit Frieda with being the first person who encouraged him to write.[5] Implausibly, they went on a carefree hitchhiking tour of Europe and lived briefly in Vienna.

But the marriage was short-lived, and the couple divorced in 1938. In the same year Frieda managed to get her elderly parents to safety in Britain, and in 1939 her brother Isidor also escaped. She supported herself—and presumably her family—first by doing domestic work, and then as a secretary. Over the years 1941–1946 she worked as a report writer for the BBC Listener Research Unit, putting in a 40-hour week at the BBC's Portland Place headquarters. By this stage she was already contemplating a career in psychology, and began work for the Medical Research Council in 1948.

It is uncertain how many of her UCL colleagues ever knew of her first marriage. Those who worked closely with her in her later years have told me they knew nothing and were surprised to learn of it. The cleverly-crafted *Times* obituary [1] skirts round the issue, as does Paul Eisler's autobiography (one can't help thinking that Eisler must have had a hand in drafting the obituary, which is unsigned).

9 Frieda and UCL Psychology

In the summer of 1937, towards the end of her brief marriage to Goldman, Frieda registered at UCL for a part-time PhD in psychology, to be completed in 1940. The following account is condensed from a 23-page dossier of Frieda's student file, kindly supplied by the UCL Records Office. She began attending a wide range of taught courses that were evidently meant to supply in accelerated form the content of a first degree (BA) in the subject. But at the end of the first year, she failed to reach the necessary standard for PhD in a qualifying examination—hardly the outcome that would be expected of one who was already qualified in psychology. She transferred instead to an MA course (for which the entry requirements were less demanding), with the intention of using that as a springboard for another attempt at the PhD qualifying exam in June 1939. But then, as UCL's Arts Tutor later wrote to the central University authorities, 'she was obliged to discontinue her course for financial reasons, and during the war, has been engaged on other work'.

In the autumn of 1945 she re-entered, on a track that would nominally have led to the MA in 1947, but again with the intention of transferring to a PhD. This doesn't seem to have worked out, since two years passed, but she did not gain an MA, or attempt the exam, and re-entered yet again in the summer of 1947 to take an examination in December that year (this looks like a postponement or a re-sit).

Her prospective PhD supervisor, J. C. Flügel (1874–1955), was all for bending the rules and submitted that she 'has done work of a high quality and has generally shown ability, understanding and application'. But the central University authorities were inflexible. They now claimed to have no trace of Frieda's 1937 entry form, and decided that even her post-war attendance could not count retrospectively towards a PhD because she still had not passed the qualifying exam. In effect she was being required to start again from the beginning. By 1949 she must have decided that enough was enough. Besides, the research that was to have formed the thesis had by now been largely completed and was beginning to appear in published form. In July 1949, twelve years after her first registration, the university authorities wrote to Frieda acknowledging a notification that she no longer wished to proceed with the PhD—and refunding the pathetic sum of five guineas.

Levelt, in an excellent account of Frieda and her London context (16: 552), accurately reports that Frieda did further training in psychology after her arrival in England. He wishes to see this as a continuation of her work with Bühler, and says that she 'completed her interrupted study of psychology'. But ironically she 'completed' nothing, and her postgraduate student record is a painful litany of 'interrupted study'. Of course 1938, when the crucial qualifying exam was failed at the outset, coincides in time with her divorce from Goldman and the effort to get her parents safely out of Austria.

Since her Vienna Dr.Phil was in literature, and her UCL studies were never completed, we are led to the extraordinary realisation that she had no formal qualification in psychology at all.

10 Conclusion

Regrettably, we have to conclude that Frieda constitutes only a tenuous link between the Linguistic Circle of Prague and the UCL Phonetics Department. Maybe Bühler did sow a seed of inspiration that eventually guided her into a career in psychology—but there's no overt indication of it, and it's strange that she doesn't even mention him. The completion of her Vienna Dr.Phil was not followed by a period of research and publication in psychology. And when she did return to psychology, she was not at first concerned with language. Her earliest publications, based on the unfinished thesis she had planned with Flügel, investigate the role of breastfeeding in character formation, and are designed as an empirical test of some claims of Freudian psychoanalytic theory. Even the work on interviews, when it started, was concerned with the interview as a sort of diagnostic tool rather than a specimen of spoken language. Frieda's work on spontaneous speech developed almost twenty years after her arrival in Britain, in the context of its time and in response to the opportunities that came along.

She was not a Viennese psycholinguist who emigrated to Britain, but a Viennese emigrant to Britain who (eventually) became a psycholinguist.

11 Acknowledgments

Very many friends and colleagues have kindly supplied information, but I am particularly grateful to Michele Cohen, who worked as Frieda's research assistant in the 1960s, and to Emma Goldman, one of three children from Willy Goldman's third marriage. I am grateful also to the editors and two anonymous reviewers.

References

[1] ANON. Prof. Frieda Goldman-Eisler. *The Times*. London. 27 January, 1982.

[2] AYER, A. J., J. B. S. HALDANE & C. CHERRY. *Studies in communication.* Secker & Warburg: London, 1955.

[3] CHAPPLE, E. D. & E. LINDEMANN. Clinical implications of measurements of interaction rates in psychiatric interviews. *Applied Anthropology*. Society for Applied Anthropology 1(2). pp. 1–11, 1942.

[4] EISLER, PAUL. *My life with the printed circuit*. (Ed.) M. E. W. Williams. Bethlehem: Lehigh University Press, 1989.

[5] CUNNINGHAM, V. Willy Goldman. *The Guardian*. London. 6 July, 2009.

[6] FRY, D. B. The experimental study of speech. In [2], pp. 147–167, 1955.

[7] FRY, D. B. & P. B. DENES. The solution of some fundamental problems in mechanical speech recognition. *Language and Speech* 1(1), pp. 35–58, 1958.

[8] GARLAND, H. & M. GARLAND. *The Oxford Companion to German Literature*. 3rd edn. Oxford University Press, 2005.

[9] GOLDMAN, EMMA. My father, Willy Goldman. *London Magazine* 33(9), pp. 81–86, 1994.

[10] GOLDMAN-EISLER, FRIEDA. The measurement of time sequences in conversational behaviour. *British Journal of Psychology. General Section*. John Wiley & Sons, Ltd 42(4), pp. 355–362, 1951.

[11] GOLDMAN-EISLER, FRIEDA. Speech-breathing activity—a measure of tension and affect in interviews. *British Journal of Psychology. General Section,* pp. 52–63, 1955.

[12] GOLDMAN-EISLER, FRIEDA. Speech production and language statistics. *Nature* 180(4600), pp. 1497–1497, 1957.

[13] GOLDMAN-EISLER, FRIEDA. Speech production and the predictability of words in context. *Quarterly Journal of Experimental Psychology* 10(2), pp. 96–106, 1958.
[14] GOLDMAN-EISLER, FRIEDA. *Psycholinguistics: experiments in spontaneous speech.* London: Academic Press, 1968.
[15] HINTON, J. *The Mass observers: a history, 1937–1949.* Oxford: Oxford University Press, 2013.
[16] LENNEBERG, E. H. *New directions for the study of language.* Cambridge, Mass: MIT Press, 1966.
[17] LEVELT, W. J. M. *A history of psycholinguistics: The pre-Chomskyan era.* Oxford: Oxford University Press, 2013.
[18] MAAS, U. *Verfolgung und Auswanderung deutschsprachiger Sprachforscher 1933–1945.* 2 vols. Tübingen: Stauffenburg Verlag, 2010.
[19] O'CONNELL, D. C. & S. KOWAL. Pausology. In S. Y. Sedelow & W. A. Sedelow (eds.), *Computers in language research 2*, pp. 221–301. Berlin: De Gruyter Mouton, 1983.
[20] PIERCE, J. The early days of information theory. *IEEE Transactions on Information Theory* 19(1), pp. 3–8, 1973.
[21] PROST, E. Emigration und Exil österreichlicher Wissenschaftlerinnen. In [25], pp. 444–470, 1987.
[22] RÖDER, WERNER, ET AL. *Biographisches Handbuch der deutschsprachigen Emigration nach 1933.* München: K G Saur, 1980.
[23] SHANNON, C. E. A Mathematical Theory of Communication. *Bell System Technical Journal* 27, pp. 379–423. Reprinted in [24], 1948.
[24] SHANNON, C. E & W WEAVER. *The mathematical theory of communication.* Urbana: University of Illinois Press, 1949.
[25] STADLER, F. & Institut für Wissenschaft und Kunst (Vienna, Austria) (eds.). *Vertriebene Vernunft* (Veröffentlichungen Des Ludwig Boltzmann-Institutes Für Geschichte Der Gesellschaftswissenschaften. Sonderband 2-). 1. Aufl. Wien: Jugend und Volk, 1987.

---

[1] She is rather familiarly referred to in the remainder of this paper as 'Frieda' (which is how she was known to close colleagues and her research students). This seems preferable to the formal designation 'Goldman-Eisler', which was never more than a pen-name anyway. The name Goldman-Eisler appears both with and without the hyphen, apparently at random. In this paper the hyphenated form is used throughout.

[2] The particulars of Frieda's UCL career have been taken from the annual College *Calendars*, which list the names and roles of all academic staff year by year. It should be noted that the title 'Lecturer' does not necessarily imply that Frieda offered any regular lecture courses—and indeed there are none listed against her name in the *Calendars*. Her teaching duties were instead in the supervision of research students.

[3] Johann Nepomuk Nestroy (1801–1862) was an Austrian singer, actor and playwright. An ebullient and charismatic anti-establishment figure, he was a thorn in the side of the conservative authorities. See [8].

[4] This section draws on numerous previously under-utilised sources, but particularly [22], which has the advantage over [18] of having been prepared and published during Frieda's lifetime. Abbreviated codes in the entry for Frieda [22: 251] indicate that the information was supplied in a questionnaire returned by Frieda herself, probably in 1975.

[5] Goldman published his first short story in 1935 [9]. His best-known work is *East End My Cradle*, on which he was working during the marriage to Frieda. It was published in 1940. When it was re-issued in 1988 Goldman added a dedication 'In loving memory of … Professor Frieda Goldman-Eisler (1907–1982), Founder and Director of the Psycholinguistics Department of the University of London'. In his partly autobiographical novel *The Light in the Dust* (1944) the fondly-portrayed character Hedi seems to be in part modelled on Frieda.

Between science and art: the speech melodies of Leoš Janáček

Markéta Vlková
Masaryk University, Faculty of Arts, Institute of Musicology
marketa.vlkova@email.cz

Abstract

When discussing Leoš Janáček – whether in the field of musicology or among the general public – the topic of speech melodies often arises. What provokes interest is not only the composer's peculiar activity of listening to human speech and recording its intonation in standard musical notation but also the musical features of such melodies. However, the view of speech melodies has changed over time, and even today, there is no general agreement on how to interpret them. The first generation of Janáček scholars, encouraged by the composer himself, considered these notated extracts to be scientific documents of real sounds which could facilitate the research of spoken language. This view was later criticised by younger musicologists, who emphasized the high level of stylization and subjective qualities of speech melodies, such as the author's favourite melodic patterns. A third, alternative view has been provided by the writer Milan Kundera. In his essay *Můj Janáček*, he describes Janáček's activity as a ritual of everydayness, through which the composer searches for the "vanished present" and the "melodic truth". This paper provides an overview of Janáček's concept of speech melodies against the background of all three abovementioned approaches. Discussing different categories of speech melodies – namely, *normal*, *excited*, and *lyrical* ones – it demonstrates Janáček's scientific ambitions and striving for precision as well as his artistic creativity and a tendency to stylise. Also, especially in light of Kundera's concept, speech melodies reveal themselves as tools to capture the elusive present.

Introduction

When the Czech composer Leoš Janáček is mentioned, it often evokes an image of a vital elderly man, running through Brno streets and parks with a notebook in his hand and jotting down fragments of overheard speech, birds singing or city noises in standard musical notation. His peculiar activity has always drawn scholarly interest. Constituting such a large amount of Janáček's endeavour, it influenced both his compositional style and literary writings, and it can therefore partially hint at his aesthetic intentions. However, when it comes to interpretation or analysis of Janáček's speech melodies, most musicologists remain slightly clueless, confronted with Janáček's obscure language and thousands of short extracts of notated speech available in the Janáček Archive (see the next section).

This paper aims to briefly present Janáček's concept of speech melodies. Relating to fundamental approaches in Janáček research, it stresses both scientific and artistic aspects of the composer's activity. It does not attempt to illuminate any new details of Janáček's theory but rather to outline it as a whole to a non-musicological audience.

1 Speech melodies

The term "nápěvky mluvy", traditionally translated as "speech melodies", was coined by Janáček himself in his study *O hudební stránce národních písní moravských* (On the Music Aspect of Moravian National Folksongs), published in 1901 as a preface to the folksong collection on which he collaborated with František Bartoš [1: 111–272]. He generally used the term not only in relation to human speech but also for the notated sounds of nature (birds, bubbling water) and occasionally inanimate objects (bells, noise from the street); he notated even the ringing in his ear. While the records of inanimate sound sources could have been collected out of curiosity or habit, the notation of animal sounds and human speech was a substantial part of Janáček's scientific endeavour. For the sake of thematic unity, this paper leaves aside the melodies of inanimate sounds or animals, restricting its focus solely on human speech.

Janáček's interest in speech arose in the 1880s. During this time, he repeatedly commented on the relationship between speech and music in his published newspaper articles. For instance, in the critique *Národní divadlo v Brně* (The National Theatre in Brno), released in his own periodical *Hudební listy* in 1886, he commented on the musico-literary relation within the form of melodrama and also observed the intonation of actors' speech on stage, taking psychological perspective into consideration [2: 96–98]. Janáček started to systematically record, collect and reflect speech melodies in the summer of 1897 [3: 479] and continued in this activity until his death in 1928. He regularly wrote down notated extracts of overheard speech, complementing many of them with information on external circumstances (time, place, situation) or the speaker's state of mind.

Crucial sources of the speech melodies are his 56 preserved notebooks, deposited in the Janáček Archive of Moravian Museum's Music History Department. However, Janáček's notated speech melodies appear also in other sources, which form two main categories: handwritten notes in other documents (such as correspondence or books from Janáček's library) and printed versions, published as part of Janáček's articles during his lifetime. The number of preserved speech melodies is estimated at more than 6 000 [4].

Although Janáček's speech melodies are a popular topic in the field of musicology, the view of them has changed over time and even today, there is no general agreement on how to interpret them. The developement of scholarly approaches to this topic was summarized by Alena Borková [5] and later by Veronika Vejvodová and Miloš Štědroň [6]; this paper builds on the latter account, preserving the categorisation outlined by Štědroň and Vejvodová. The first generation of Janáček scholars, encouraged by the composer himself, considered these notated extracts to be scientific documents of real sounds which could facilitate the research of spoken language. This view was later criticised by younger musicologists, who emphasized the high level of stylization and subjective qualities of speech melodies, such as the author's favourite melodic patterns. As John Tyrrell puts it: "For all the 'objective', 'scientific' aspects that Janáček stressed, speech melodies reflect much more of him than they did of their speakers" [3: 489]. A third, alternative view has been provided by the writer Milan Kundera. In his essay *Můj Janáček*, he describes Janáček's activity as a ritual of everydayness, through which the composer searches for the "vanished present" and the "melodic truth" [7: 30].

2 Scientific ambitions: *Normal* speech melodies

Janáček's so-called speech-melody theory is rather an unstable set of observations and assumptions than a consistent theory, but the term "theory" is traditionally used in accordance with Janáček's own texts [8: 282]. Since the composer did not write a synthetizing oeuvre on this topic, his views are distributed among numerous published or unpublished pieces of writing. Their synthesis reveals that Janáček regarded human speech from diverse perspectives: he related it to psychology, had some ambitions in the field of linguistics, and took also musical aspects into account, hoping to exploit his observations in his own compositional process.

Among several types of speech melodies that Janáček distinguished, *normal* ("normální") melodies are most closely related to his scientific interest in language. Uttered in a calm state of mind, they usually employ only one tone – or two adjacent ones – and consist of even rhythmic values. Janáček did not consider them common in everyday speech, or pay particular attention to them. It was, however, this category that formed the basis for Janáček's intended project – a sound dictionary of the Czech language. He outlines this intention in several texts, including the article *Rozhraní mluvy a zpěvu* (The Border Between Speech and Song), published in 1906 in the periodical *Hlídka*:

> "We need a book on the ordinary melodic curves of speech in order to preserve the sound of the Czech language for future generations. It would be a dictionary in notes of the living Czech language, which would contain melodic phrases for everything which the Czech language is able to express. Let me give an example: When Mrs. Úprka painted my wooden chest, she discussed the colours:
>
> She painted:
>
> 
>
>
> [...]
>
> Her conversation was quiet and peaceful; I did not want to disturb her in her work. These are what I would call normal melodic curves of speech. Here there is no annoyance, no bitterness, no anger, no gaiety, no sadness. They are pronounced almost without any alternations and accents, whether used in the morning or before going to bed. Whoever you may meet, will tell you the name of the parish in the same way. Which shows how the melodic curves of speech are general." [9: 91–92]

The composer's vision of a sound dictionary has never materialized. In his last will, he left one hundred thousand crowns to the Masaryk University, Faculty of Arts, to fund phonetic research of living speech. His wish remained unfulfilled, however, due to legal obstacles and the historical situation: before the outbreak of WWII, all materials from the faculty's Janáček Archive were transferred to the Moravian Museum, and after the closure of Czech universities in 1939, they were temporarily managed by another subject; many documents have never returned to the faculty [10: 214–218].

Despite Janáček's idealisation of normal speech melodies, he did acknowledge the influence of the speaker's subjective attributes on them. He was aware that every person has a unique way of articulating, and moreover, the speech is significantly shaped by individual dialects, which have an impact on intonation, stress or the length of vowels. Nevertheless, these individual features are independent of the speaker's emotional state, and for this reason, Janáček considered even dialectical speech melodies *normal* [11: 131].

3 From scientific psychology to stylized melody: *Excited* speech melodies

What provoked even more interest in him – this time in the realm of psychology – were the speech melodies that deviated from the normal ones, reflecting the emotional state of the speaker. Due to Janáček's unstable terminology, scholars use diverse terms for this type of speech melodies; Milena Černohorská, for instance, refers to them as *excited* ("vzrušené", also "nápěvky vzrušené mluvy") [11: 134, 137]. As Černohorská points out, it is exactly this focus on inner, subjective aspects of speech that differentiates Janáček from those linguists who examined speech intonation only with concern to the content of the speech or the word and sentence accents [11: 135]. Otakar Hostinský, for instance, focused on normal speech, exempt from affects or subjective irregularities, as he studied the objective features of declamation [12]. Janáček, on the other hand, concentrated on the emotional state of the speaker, which he believed was recognisable from the melody of speech. In this regard, his perspective resembled that of the phonetician Josef Chlumský [13]. Janáček meticulously took notes of diverse circumstances that possibly impacted the final speech melody in terms of intonation, rhythm, register and speed of delivery. He viewed the speech melodies as signs of a person's inner state, which is formed by external conditions, and therefore inseparable from their original environment from which they stem, i.e. from both the external situation and the speaker's consciousness. As he explains in a *feuilleton*[1] from the periodical *Lidové noviny* entitled *Moravany! Morawaan!*:

> "Is it possible to retain in one's mind a speech melody and pull back a little the curtain behind which it is being born? It issues from the lips as an image embossed on a gold coin emerges from the mint. We have to consider the whole coin: *you cannot select only the relief of the image*. This melody, its surface and edges, are of one metal: this is how speech melody is joined together with the contents of our consciousness. This is how it is moulded together with the reflection of the speaker's inner life, and the reflection of the environment in which it is spoken." [14: 42–43]

Janáček indeed respected the link between a speech melody and its original environment and preserved it even in his *feuilletons*, which are otherwise often stylized and attest to the author's literary ambitions [15].

What has often led to misunderstanding – among both Janáček's contemporaries and younger musicologists – was his view of notation. Naturally, there is a significant gap between music and a "melody" of speech, especially when the sound is recorded by means of standard musical notation, which restricts the intonation to pitches of a narrow range within a mere chromatic scale. Moreover, a conspicuously large number of speech melodies consist of Janáček's favourite musical intervals – seconds, fourths and fifths [16: 133–135]. The rhythm might seem much more precise in comparison: many notated extracts feature complicated rhythmic patterns and diverse tone lengths, thus evoking a sense of haphazardness, which may approach the nature of a non-stylized human speech. Such sense of objectivity is strengthened by Janáček's precise inscriptions of the melodies' duration. Especially after 1922, when he – inspired by the German psychologist Wilhelm Wundt – started using the Hipp's chronoscope to measure time, the inscriptions become exceptionally specific, recorded to the nearest hundredth second. However, as John Tyrrell points out, even the rhythm does not fully reflect reality, as the chronoscope was not portable, and Janáček had to keep the sound in his memory before measuring it at home [3: 478].

Admittedly, such notation inaccuracies can be interpreted as evidence of Janáček's creativity and aesthetic preferences, and in the case of intervals, it seems plausible. Nevertheless, Janáček himself considered conventional notation unsatisfactory, as it approaches the original sound of speech only roughly. In the article *Rozhraní hudby a zpěvu*, he reflects the trivialisation and "hardening" of speech melodies when written in notation, reminding also that speech melodies are not intended to be sung [2: 348]. Therefore, it would be a misinterpretation to deny the scientific nature of Janáček's records only on based on (self-evident) notation imprecision or insufficiency.

[1] Many of Janáček's newspaper articles – especially those published in *Lidové noviny* – are traditionally referred to as *feuilletons*. The literary genre of the *feuilleton* and the style and content of Janáček's *feuilletons* are characterised by Tiina Vainiomäki [8: 129–133], who describes these texts as "small snapshot-like *belletristique* articles" [8: 129].

4 Artistic stylization: *Lyrical* speech melodies

The subjective nature of Janáček's observation emerges in connection with another category of speech melodies, for which Černohorská used the term *lyrical* ("lyrické") [11: 141]. When studying such melodies, Janáček does not concentrate on the speaker's current affect revealed in the utterance at a particular moment, but rather attempts to capture a general mood of the speech. He usually examines several lyrical speech melodies together, trying to find their underlying key which unifies them – an *emotional scale* ("stupnice citová") [17: 432]. Such search for a key as a framework for several speech melodies was a conscious process, reflected by Janáček himself in his writings, such as in *Rytá slova* (Words Engraved) from the newspaper *Lidové noviny* [2: 482–484].

Using music theory vocabulary, he frequently described this kind of speech almost as if it was a musical composition. After attending the linguist Josef Zubatý's lecture in October 1925, Janáček published an eponymous *feuilleton* in *Lidové noviny* where he described Zubatý's intonation as ranging from f^1 to B♭ with the timbre of viola d'amour [2: 563]. This tendency manifested itself even stronger in the aforementioned article *Rytá slova*, which comprised numerous speech melodies of Tomáš Garrigue Masaryk, the first president of Czechoslovakia [2: 482–484]. Summarizing Masaryk's two-day lecture, Janáček even mentions different keys of the president's speech on the two respective days: while during the first day he found the speech dominated by A-flat minor without modulations, in the next day's speech he heard Hypolydian D with a changing second (E–E-sharp) and fourth (G–G-sharp), which reminded him of the sun without mists. At the end of the *feuilleton*, he even put down the two chords of the respective keys:

Figure 1. Two underlying chords of Masaryk's speech in the *feuilleton Rytá slova* [2: 484]

Rytá slova is not the only *feuilleton* of Janáček's that balances on the borderline between speech examination and music composition. In several other writings, the composer writes down a speech melody and later repeats it with harmonized accompaniment. For instance, in the *feuilleton Počátek románu* (The Beginning of a Romance), he captures a dialogue of two ladies at the train station in Brno. One of them says: "We will stand here, and I know he's not coming!" The other woman responds: "It doesn't matter!" [2: 495].

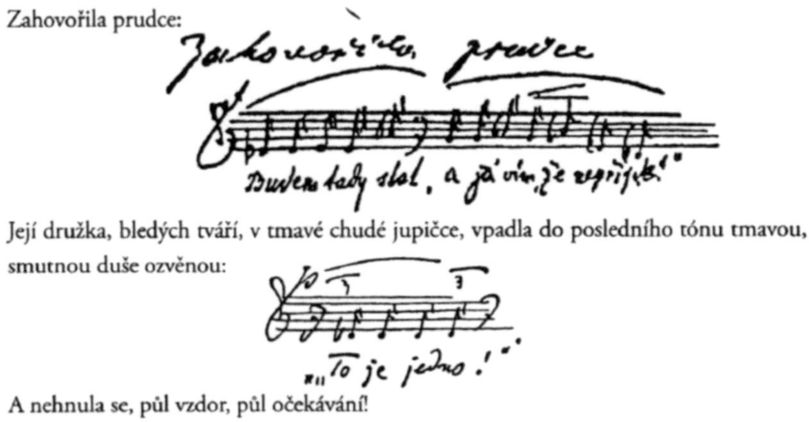

Figure 2. Extract from the *feuilleton Počátek románu* [2: 495]

Then he cites both melodies (albeit slightly changed and divided into measures), this time with harmonization, which attributes a particular key to them.

Figure 3. Harmonized speech melodies in the *feuilleton Počátek románu* [2: 495]

Although there is undeniably a certain level of stylization, the depicted process suggests that the idea of a key stems from the observation and does not precede it. In his study on naturalism, Janáček himself confirms that he does not have an idea of a particular chord before he jots down speech melodies, but – on the contrary – the chord or key arises from the speech extracts as their result [18: 174; English translation: 19: 297].

5 Speech melodies as pictures of the present

In his essay *Můj Janáček* [7] (the first section of which – *A la Recherche du Présent Perdu* – is part of *Les testaments trahis*, in English *The Testaments Betrayed*), Milan Kundera provides an alternative interpretation of Janáček's speech melodies. He describes Janáček's activity as a ritual of everydayness, through which the composer searches for the elusive present and the truth of expression:

> "The search for the vanished present; the search for the melodic truth of a moment; the wish to surprise and capture the fleeting truth; the wish to plumb by that means the mystery of the immediate reality constantly deserting our lives, which thereby becomes the thing we know least about. This, I think, is the ontological import of Janáček's studies of spoken language and, perhaps, the ontological import of all his music." [20: 138]

Janáček indirectly confirms this view in the *feuilleton Loni a letos* (This Year and Last) from the periodical *Hlídka*: "The melody of speech is a truthful transient musical characterization of a person; it is his soul and encompasses his entire being in a photographic instant" [21: 246]. The metaphor of a photograph then recurs in the composer's later texts on speech melodies.

When the melodies appear in Janáček's literary articles, notation – alongside many literary devices – becomes a crucial tool for expressing time. First, it unites "narrating time" and "narrated time": encountering notation, a reader is forced to slow down the process of reading and "play" the speech or other sound at a similar speed as it originally sounded. This effect becomes particularly obvious in the writings where speech melodies dominate, such as the abovementioned *Rytá slova*. Second, it provides a sense of the present even when used within a description of the past, namely in citation of a person who quotes someone else or in an indirect speech. On another level, the present is expressed by the haphazard order of ideas in Janáček's texts – the more the hierarchy of motives and coherence are neglected, the more a particular moment arises in its realness, captured in the way it is immediately perceived. This is, however, related more to Janáček's literary style than to his views of speech melody.

Probably the most poignant examples of Janáček's attempt to capture the present are the speech melodies of his dying twenty-year-old daughter Olga. In February 1903, during the last days before her death, Janáček recorded some of her sentences and words.

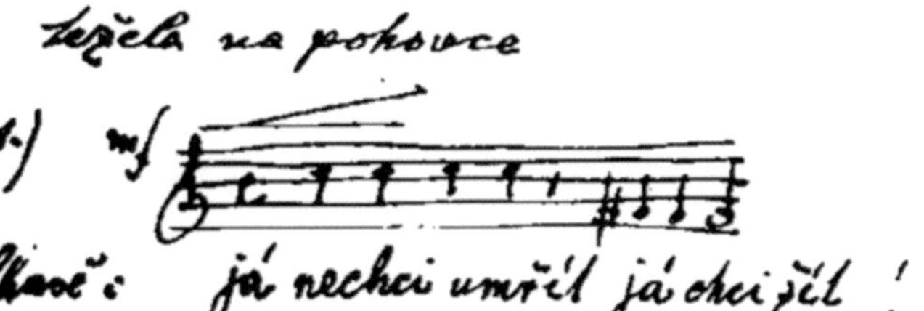

Figure 4. Speech melody of Janáček's daughter ["lying on the settee, sobbing: *I don't want to die, I want to live!*"]. Reproduced and translated in [8: 179].

Figure 5. Speech melody of Janáček's daughter [last sights: *"ah-h"*; God be with you, my darling]. Reproduced in [8: 179], translated in [22: 258].

As Paul Christiansen discusses, it might seem even ghoulish that in such a situation, Janáček paid attention to his daughter's voice intonation and musical attributes of her speech [22: 255]. However, her speech melodies – unlike most other ones – were neither scientific material nor creative sources to Janáček: they were simply the composer's way of capturing the present moment and immortalizing the memory of his child. Kundera's concept of the vanished present perfectly illuminates such intention.

Conclusion

Leoš Janáček's speech melodies represent an attractive yet challenging research topic. Their interpretation is hampered by the fragmentation of the composer's theory and by his specific, almost untranslatable language, which practically confines the research to scholars with excellent command of Czech. The disunity of Janáček's explanations has been mirrored by the diversity of scholarly approaches, emphasizing either the scientific or the artistic aspects of the composer's activity.

This paper demonstrated selected features of Janáček's speech-melody theory, relating them to three ways of interpretation. Each perspective thus illuminated different aspects of such a complex topic. Naturally, Janáček's speech melodies could be examined from many other perspectives and represent an ideal object for interdisciplinary research. Hopefully, presenting this topic outside of the musicological audience might also contribute to wider interdisciplinary collaboration and deeper understanding of this phenomenon.

References

[1] JANÁČEK, L.: *Folkloristické dílo (1875–1928)*, I/3–1. Eds. J. Procházková, M. Toncrová & J. Vysloužil. Brno: Editio Janáček, 2009.

[2] JANÁČEK, L.: *Literární dílo (1875–1928)*, I/1–1. Eds. T. Straková & E. Drlíková. Brno: Editio Janáček, 2003.

[3] TYRRELL, J.: *Janáček: Years of a Life. Volume I (1854–1914)*. London: Faber and Faber, 2006.

[4] ŠTĚDROŇ, M.: O projektu. In: *Nápěvky mluvy* [online database, accessed 20 February 2021], http://napevkymluvy.cz/index3.php?info=list.

[5] BORKOVÁ, A.: Pohled na Janáčkovu nápěvkovou teorii očima české muzikologické literatury do 60. let 20. století se zmínkou o okolnostech jejícho vzniku. In: J. HALÍŘ, L. PELÍŠKOVÁ & P. SÝKORA (Eds.): *Fenomén Janáček včera a dnes*. Brno: Editio Conservatorium Brunense, vol. I, pp. 185–191, 2006.

[6] VEJVODOVÁ, V. & ŠTĚDROŇ, M.: O databázi nápěvků mluvy Leoše Janáčka – první poznatky a zkušenosti z projektu. *Opus musicum* 39(3), pp. 19–20, 2007.

[7] KUNDERA, M.: *Můj Janáček*. Brno: Atlantis, 2004.

[8] VAINIOMÄKI, T. *The Musical Realism of Leoš Janáček. From Speech Melodies to a Theory of Composition* [online]. Doctoral dissertation, University of Helsinki, Faculty of Arts, Department of Philosophy, History, Culture and Art Studies, 2012. Accessible from: https://helda.helsinki.fi/bitstream/handle/10138/36087/themusic.pdf [accessed 2021-05-16].

[9] JANÁČEK, L.: *Leoš Janáček – Letters and Reminiscences*. Ed. B. Štědroň. Translation by G. Thomsen. Prague: Artia, 1955.

[10] MACEK, P. Leoš Janáček a jeho odkaz Filozofické fakultě Masarykovy univerzity. In: J. HALÍŘ, L. PELÍŠKOVÁ & P. SÝKORA (Eds.): *Fenomén Janáček včera a dnes*. Brno: Editio Conservatorium Brunense, vol. I, pp. 213–218, 2006.

[11] ČERNOHORSKÁ, M.: Nápěvková teorie Leoše Janáčka. *Časopis moravského muzea* XLIII, pp. 129–144, 1958.

[12] HOSTINSKÝ, O.: *O české deklamaci hudební*. Prague: F. A. Urbánek, 1886.

[13] CHLUMSKÝ, J.: *Česká kvantita, melodie a přízvuk*. Prague: Česká akademie věd a umění, 1928.

[14] JANÁČEK, L.: *Janáček's uncollected essays on music*. Ed. M. Zemanová. London – New York: Marion Boyars, 1989.

[15] VLKOVÁ, M. *Hudba v Janáčkových fejetonech z Lidových novin* [online]. Master thesis. Masaryk University, Faculty of Arts, 2018. Accessible from: https://is.muni.cz/th/mfr6p [accessed 2021-05-16].

[16] ŠTĚDROŇ, M.: *Leoš Janáček a hudba 20. století. Paralely, sondy, dokumenty*. Brno: Nadace Universitas Masarykiana – Edice Scientia, 1998.

[17] JANÁČEK, L.: *Teoretické dílo (1877–1927)*, I/2–1. Eds. L. Faltus, E. Drlíková, S. Přibáňová & J. Zahrádka. Brno: Editio Janáček, 2007.

[18] JANÁČEK, L.: *Literární dílo (1875–1928)*, I/1–2. Eds. T. Straková & E. Drlíková. Brno: Editio Janáček, 2003.

[19] JANÁČEK, L.: Janáček on Naturalism (1924–1925). Ed. M. Štědroň. In: M. BECKERMAN (Ed.): *Janáček and His World*. Translation by V. Firkušný-Callegari and T. Firkušný. Princeton: Princeton University Press, pp. 287–304, 2011.

[20] KUNDERA, M.: *Testaments Betrayed*. Translation by Linda Asher. London: Faber and Faber, 1996.

[21] JANÁČEK, L.: Last Year and this Year (1905). In: M. BECKERMAN (Ed.): *Janáček and His World*. Translation by V. Firkušný-Callegari and T. Firkušný, Princeton: Princeton University Press, pp. 239–253, 2003.

[22] CHRISTIANSEN, P.: The Meaning of Speech Melody for Leoš Janáček. *Journal of Musicological Research* 23(3–4), pp. 241–263, 2004.

ON PHONETIC TRANSCRIPTION: ACHIEVEMENTS OF CREATIVE MINDS

Mária Gósy

Hungarian Research Centre for Linguistics and Applied Linguistics and Phonetics Department of ELTE University, Budapest, Hungary
gosy.maria@nytud.hu

Abstract: The attempts to create a phonetic writing system have had a relatively long history of over 500 years. Various inventors worked on establishing a transcription system that reflects the articulation of a language. Beside the well-known names in this field, there are a few Hungarian scientists who contributed to the development of transcription without, alas, leaving a trace for the international body of knowledge. They created both analphabetic and alphabetic writing systems, but the majority of their inventions remained unknown, unused and ineffectual even in Hungary. Looking at these "phonetic scripts", however, the successors should acknowledge the attempts, thoughts, creativity and results of these people that in a way enrich the history of phonetics. This study introduces these scientists, and places their inventions into a broader context of the history of the transcription of speech.

1 Introduction

The traditional writing systems of most languages may originally have been some kind of transcriptions of speech, but that situation has changed across the centuries [8], [10], [15]. The demand to visualize the language-specific pronunciation properties of various speech sounds in writing emerged and resulted in diverse ways of resolving this task during the past centuries. The obvious start was the recognition that transcriptions are entirely different from orthographic representations, in that the former need to employ a kind of phonetic notation to refer to a sound unambiguously (whereas the latter may have other, partly contradictory motivations). Both the reasons for creating such systems and the scholars' professions were interestingly different.

The creators of phonetic writing systems – beside linguists and phoneticians – had various professions that were connected to or distant from the idea of transcription, such as physicians, speech therapists, a schoolmaster, a lawyer, a knight, a teacher, a diplomat, an ophthalmologist, a trader, a teacher of the English language who had developed the most widely used system of shorthand, a traveler, a parson, and a physiologist. The motivations for creating a phonetic transcription system ranged from pure linguistic, phonetic interest across developing a system that can be used in the description of less known languages, reducing class barriers by establishing a standard form of pronunciation, and learning foreign languages, up to helping to teach written language to children and particularly to teach speaking to hearing impaired and deaf people.

Although the main trends of the development of phonetic transcription have been well described in the literature (e.g., [8], [10], [15]), there are interesting and important details that – for some reason – do not appear in the summaries.

The goal of the present paper is twofold. I intend to introduce the transcription systems of some Hungarian inventors that seem to be unknown to the community interested in the history of phonetics. In addition, I would like to place their inventions into a broader international context of the development of transcription of speech. I am of the opinion that their achievements are worth studying and introducing since they are important parts of the international heritage of phonetic writings. I will focus mainly on the time between the beginning of the 19th century and the beginning of the 20th century. I will discuss a few writing systems that may have influenced the Hungarian scientists. The overview of those inventors' findings helps to interpret the paths Hungarian scientists followed. I admit that the selection of the international heritage is entirely subjective while I intend to show every important detail of the historical facts of the topic in relation to Hungarian scientists to be listed. The history of Hungary may be important to properly appreciate the context in which Hungarian scientists worked (see [17]). I will introduce all attempts by Hungarian inventors irrespective of their motivations, mental backgrounds and the language they used in their written studies.

2 International steps toward a transcription system

The first attempts to establish a transcription system go back to the 16th century. Sir Thomas Smith (1513–1577) introduced the diacritic [¨] to indicate long vowels as in [ë] and [ä]. Kemp's detailed survey [10] excellently shows the history of transcription. During the following centuries various trends could be seen in attempting to create transcription systems. Broadly, two trends can be distinguished, those of alphabetic and analphabetic systems; however, neither of them seems to be homogeneous. Analphabetic writings tried to visualize the articulation gestures in writing to create a "universal" (language-independent) alphabet. These systems broke with traditional letters and tried to use entirely new symbols. The other trend used mostly Latin letters amending them to approximate the corresponding speech sounds.

The beginnings of the analphabetic notation systems were initiated by Dutch and English inventors, Lambert ten Kate (1674–1731) and Erasmus Darwin (1731–1802), as well as Thomas Wright Hill (1763–1851), an English schoolmaster (see e.g., [10]). Hill represented the active and passive articulators as well as the articulation gestures by digits and lines in his writing system (e.g., a bilabial articulation is number 1, a labiodental articulation is number 2, while the state of the glottis is shown by the shape of the line between the digits). One of the most often cited scholars was the German physician and physiologist Ernst Wilhelm Brücke (1819–1892). His conception was that a phonetic writing was to be universal, that is, appropriate to reflect the articulation (the subcomponents) of the segments of any language. He provided transcriptions of read texts based on the production of standard German and 3 German dialects as well as 19 other languages including Hungarian [5: 52-65]. The Hungarian speaker might have spoken a dialect of the Somogy area (Figure 1).

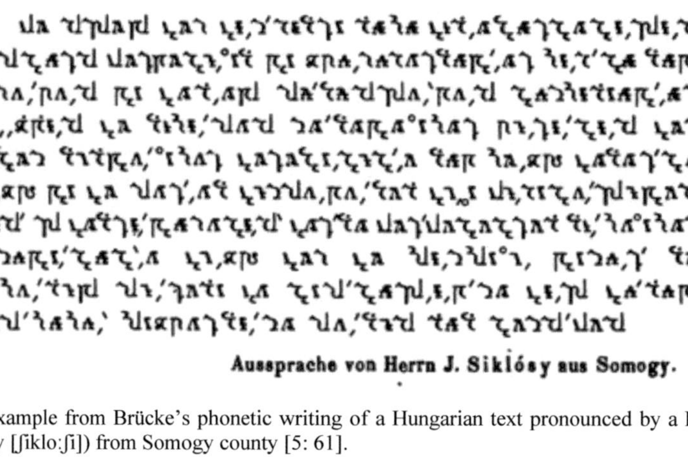

Figure 1. Example from Brücke's phonetic writing of a Hungarian text pronounced by a Hungarian man (Siklósy [ʃiklo:ʃi]) from Somogy county [5: 61].

He and his followers tried to apply the principle that the symbols were to be based on the articulation organs and their successive motions. These phonetic symbols should be some sort of "articulation recipes", exact prescriptions of how to articulate the given speech sound. Interestingly, Brücke knew the system of the first Hungarian inventor, Sándor Kyss (see later), and discussed and criticized the Hungarian inventor's attempts in his work [4].

Another scientist, Moritz Thausing (1838–1884) created a phonetic writing that might have been influenced by musical notation [20]. He placed his various symbols whose note heads were different according to their form and direction into a 4-line system (Figure 2). He was convinced that this transcription was suitable to visualize the "natural system" of the speech sounds that he called a "linguistic alphabet".

Figure 2. Examples from Thausing's phonetic writing. The bottom part of the figure shows his own transcription of his book's title: *Das natürliche Lautsystem der menschlichen Sprache* [20: 132].

The Scottish-born Alexander Melville Bell (1819–1905) followed Brücke's theory in creating his transcription system [3]. Bell's principle in developing his phonetic writing (known as "visible speech") was that it was meant to visualize the process of articulation. His letters were simple and provided some scheme of the articulation gestures (e.g., various forms of lines that represented the articulation gestures and the connection between segments). Bell's system contained 61 different symbols. He applied his transcription to various languages while he wanted to help hearing impaired and deaf people (Figure 3).

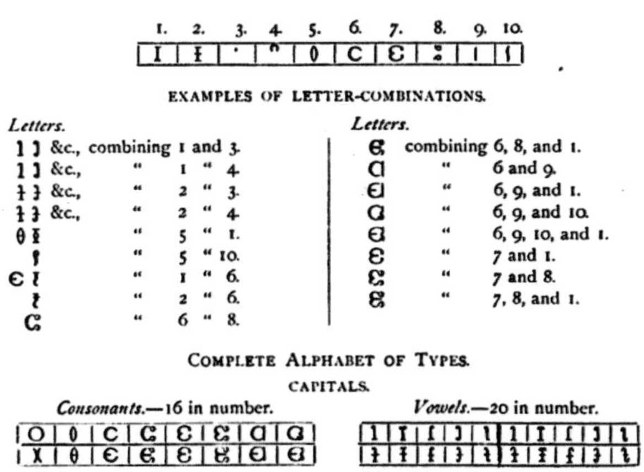

Figure 3. Fragment from Alexander Melville Bell's visible speech [3: 36].

The Danish linguist Otto Jespersen (1860–1943) contributed to various fields of linguistics including phonetics that was one of his greatest areas of interest [9]. His analphabetic transcription is unique for several reasons. The author intended to define each articulation gesture of a speech sound properly using various symbols. Jespersen's transcription system contained a combination of Roman letters, Greek letters, numerals, italics, and subscript as well as superscript letters [9]. The conception and desire of reflecting the subcomponents of the articulation in writing can be seen in his images. The transcription of the consonant [d] and the explanation of the underlying symbols can be seen in Figure 4.

$$\beta 0^{fe}\delta 0\varepsilon 1$$

Figure 4. The North-German consonant [d] in Jespersen's phonetic writing. The symbol **β** means tongue tip, **0** means closure, **fe** represents the area of alveolar ridge/hard palate while **0fe** means that there is a closure between the alveolar ridge and the teeth, **δ** means velum and uvula while **δ0** means velic closure, **ε** means vocal cords, **1** means close stricture while **ε1** means voiced.

As can be seen, the analphabetic systems made efforts to represent the subcomponents of the segments in diverse ways. It was, however, a common "property" of the analphabetic transcriptions that they were difficult to use. Some of them showed no proper knowledge of articulation of the speech sounds and the physiology of the speech organs of the inventor, yet others caused difficulties in printing and often in continuous handwriting, as well as again others were too complicated to learn and use in practice. The analphabetic transcription systems are extremely exciting; however, they became scientific "blind alleys". They provided, in a way, reform thoughts and a lot of ideas to developing modern transcription systems.

There are a number of transcription systems that modified the Latin letters, added diacritics to them, modified/altered the letters themselves and borrowed some letters from other writing systems. The inventors of these systems made an effort to be consistent in their systems. The most important scholar of this conception was the Prussian Egyptologist and linguist, Karl Richard Lepsius (1810–1884). He intended to create a language-independent system that would be able to represent the speech sounds of any language. His transcription was intended to help the societies of missionaries in the case of languages that either had no writing systems or had non-alphabetic writings [13]. His writing system, however, was too complicated for general use, mainly due to the high number of diacritics employed. Details from his transcription are demonstrated in Figure 5.

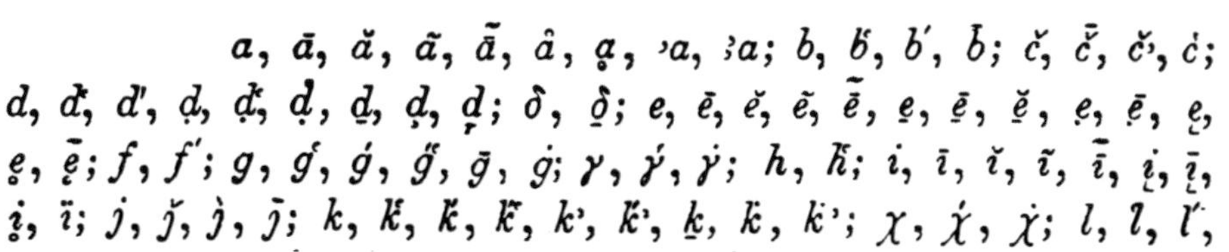

Figure 5. Examples from Lepsius's transcription system [13: 18].

In England, Pitman and Ellis were pioneers in this field. They used Latin letters and their modified versions while they considered also the printing possibilities of their symbols (see [1], [10], [12], [15]). Although the reform they advocated was never realized, their work inspired those that would be later employed to establish the alphabet of the International Phonetic Association [19]. The great English phonetician, Henry Sweet (1845–1912), initially followed Ellis but later he applied his own principles in establishing a phonetic writing (called "Narrow Romic") including the possibility of handwriting of his symbols [18]. The transcription system of the Swedish Johan August Lundell (1851–1940) was used by researchers of Swedish dialects. His writing contained mainly Latin letters, but he tried to avoid diacritics [14]. Most of these alphabets were intended to be used by phoneticians and linguists, not by the general public.

In 1922, at the meeting of the Union Académique Internationale held in Brussels, Jespersen and another scholar, Salverda de Grave intended to call for order in the transcription "chaos". The next meeting was held in Copenhagen in 1925 where 12 famous scientists attended to discuss the possible new phonetic symbols and diacritics [12], [15]. Several important decisions were made, and the basic principles were accepted at the meeting, among them the differentiation of the speech sound and the phoneme.

Although the community of linguists and also the interested scientists were aware of the need to establish a useful transcription system as early as in the 19th century, this task was not accomplished until the beginnings and development of the International Phonetic Alphabet (e.g., [1], [15]).

3 Transcription systems invented by Hungarian scholars

We have no proper knowledge about the motivations of all scientists dealing with the problem of transcription; however, we are aware of some inventors' personal motivations and goals that seem to be very similar to those of the international authors. It is extremely difficult if not impossible to define the supposed effects of foreign scholars and their attempts on Hungarian scientists. It is safe to assume that those who lived in the 19th century might have learned about some international attempts toward a specific writing system. The idea, naturally, may have emerged independently of the international impact. And again, similarly to many scientists of the world, practice itself – the needs of phoneticians, linguists, dialectologists, teachers of deaf people, etc. – could raise the demand for establishing a usable phonetic writing. There were scholars that actively reacted to these demands [21].

The very first Hungarian inventor who dealt with phonetic transcription was Sándor (Alexander) Kyss [ʃa:ndor kiʃ:], a knight and landowner as well as judge of the court of appeal in Pest-county. His work was written in Latin in 1817 and remained unpublished [11]. He defined symbols demonstrating the articulation of speech sounds. Beside letters, he used digits and other symbols in this analphabetic system. This kind of "visible speech" preceded that of Alexander Melville Bell's by about 50 years. The detail from the manuscript of Kyss (Figure 6) shows the symbols of the basic building elements of his system and their explanations. He differentiated the primary vowel (marked by a circle-like symbol), breathing, closure, breathing through mouth and nose, primary breathing, inspiration, release, closure and release, as well as sonorant(s). Sonorants were marked by concentric circles where the smaller one is black. The circle itself resembles the vowel mark, and Kyss made clear the connection between them. Unfortunately, the work of Sándor Kyss was not followed but his ideas influenced later authors like András Matúsik [ɔndra:ʃ mɔtu:ʃik] and Pál Bugát [pa:l buga:t].

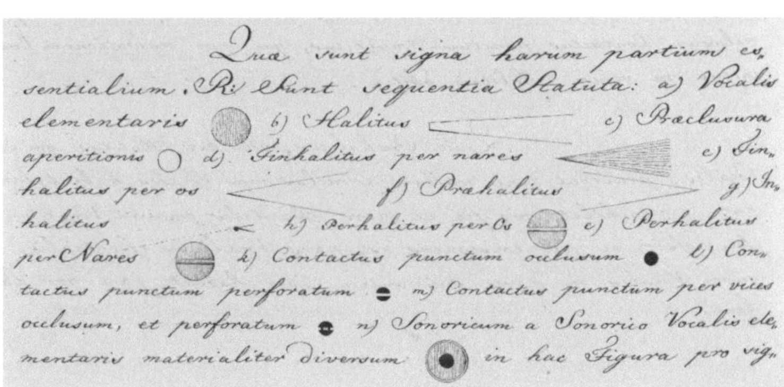

Figure 6. A detail of Kyss's manuscript with his symbols [11: 58].

András Matúsik (?–1839) was a parson teaching natural science at a secondary school in Rozsnyó (now in Slovakia), and intended to create a universal speech writing that contained Latin letters and additional symbols. He defined and followed the principle of one symbol for one speech sound. He differentiated the dialectal vowel quality of [e] from its other pronunciations. His system was able to reflect the pronunciation characteristics of any language. His book was published in 1832 [16].

Pál Bugát (1793–1865) was a physician (an ophthalmologist), a university lecturer, and member of the Hungarian Academy of Sciences. Beside his medical studies he was devoted to comparative linguistics and wrote a number of papers that remained in manuscripts. Later, he studied the Finnish and Turkish languages. Perhaps these two trends in his life were somehow interrelated when he created his interesting, analphabetic transcription. He called his work "phonorthography" [6], and it remained a manuscript. His goal was to help children to learn to write using these "speaking" symbols before teaching them to use letters. He employed a two-line system where the positions of the symbols indicated the place of articulation, like high, mid or low (see the positions of the symbols for bilabial, dental and velar stops). The symbols indicate whether the consonants they represented are voiced or voiceless: the voiced consonants have their flourish preceding the symbol while the flourish follows the symbols for the voiceless consonants (Figure 7). Bugát's transcription obviously shows his good knowledge of the articulation of speech sounds.

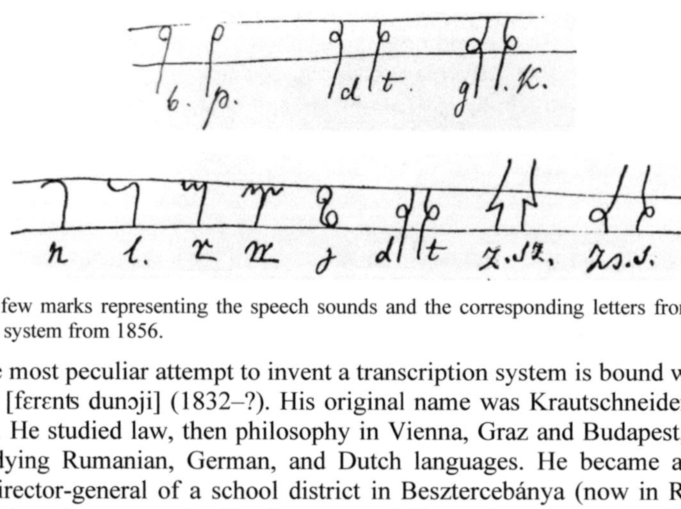

Figure 7. A few marks representing the speech sounds and the corresponding letters from Bugát's transcripition system from 1856.

Perhaps the most peculiar attempt to invent a transcription system is bound with the name Ferencz Dunay [fɛrɛnts dunɔji] (1832–?). His original name was Krautschneider and he was born in Vienna. He studied law, then philosophy in Vienna, Graz and Budapest, and was interested in studying Rumanian, German, and Dutch languages. He became a high-school teacher, later director-general of a school district in Besztercebánya (now in Romania). He could speak various languages, beside German and Hungarian, Rumanian, Dutch, Italian, French, English, Slovakian and not too well Polish and Croatian. The title of his book is *Átiró hangrajz (Phonetica transscriptoria)* [7]. The English "translation" of the Hungarian title (word by word) is something like "transcriptional sound drawing". This term is unique, a kind of neologism that nobody used later (or before). The front board of his book, shown in Figure 8, contains the usual letters and also Dunay's analphabetic transcription of the texts.

Dunay's transcription system is preceded by explanatory figures demonstrating the characteristic articulation of several speech sounds together with the appropriate elements of this writing system and by a detailed presentation of the system itself using eight tables. The author created practically two transcription systems and considered both the printing possibilities and handwriting in both cases. One of his transcription systems is analphabetic where the symbols correspond to articulations of various possible speech sounds (irrespective of the language). The other transcription system, invented by Dunay specifically for Hungarian, uses Latin letters and diacritics and it is introduced in the second part of the book.

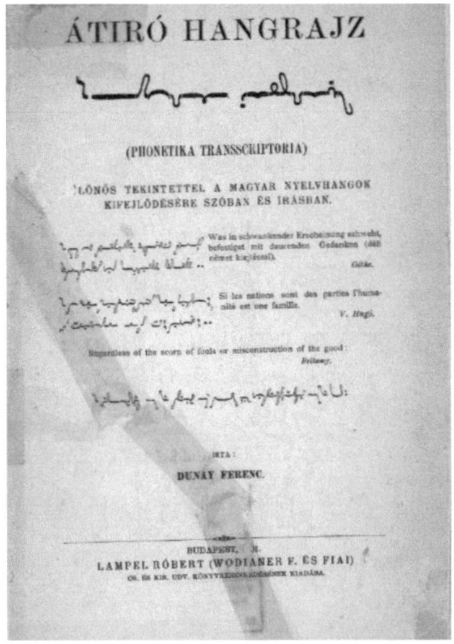

Figure 8. The front board of Dunay's book. The title (in Hungarian) and three quotations from "Göthe" (in German), V. Hugo (in French) and Bellamy (in English) are written together with their transcriptions using the author's system.

He transcribed two parts of Schiller's poem ("Columbus") spoken as on the stage and in a Southern German pronunciation shown in Figure 9. Part of a Hungarian poem was also transcribed (see [7: 17]).

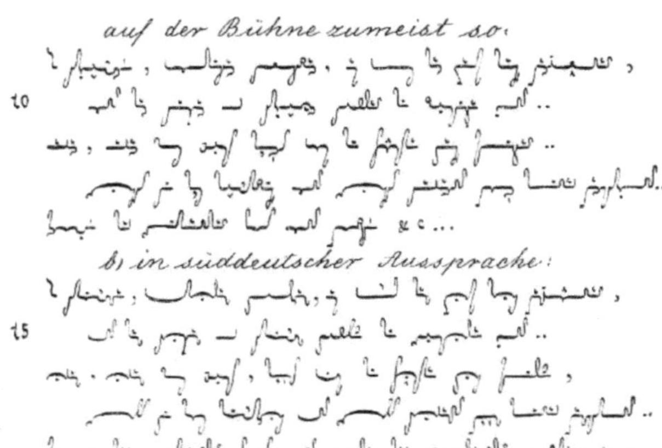

Figure 9. Two transcription parts of Schiller's poem ("Columbus") spoken as on the stage and in a Southern German pronunciation (see [7: 15]).

Dunay's book contains several examples showing the use of his analphabetic transcription system for various languages. In addition, on page 22, he demonstrated the pronunciation differences of a Latin text as read by a Hungarian, an Italian, a Polish (Figure 10) and two German speakers (two different dialects for the latter). This is the Latin text (the numbers indicate the corresponding transcriptions in the various languages): 1. Sic erim Deus dilexit mundum, ut Filium suum 2. unigenitum daret, ut omnis qui credit in eum 3. non pereat, sed habeat vitam aeternam (in English: "God loved the world so much that he gave his only-begotten son so that all who believe shall not be taken but have eternal life").

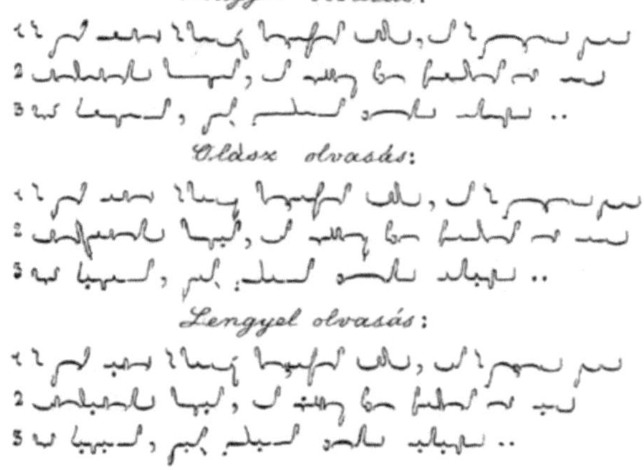

Figure 10. Transcriptions of a Latin text read in Hungarian, Italian, and Polish [7: 22].

Dunay seems to be aware of the importance of printing. Thus, he tried to create another transcription system that had both printing and handwriting versions. Figure 11 shows a few lines of a Hungarian poem (Vörösmarty, M.: ([vørøʃmɔrti]): *Zalán futása* "The Flight of Zalán") transcribed by this system. (The handwriting shows specific ascending lines which is supposed to be the author's usual handwriting.)

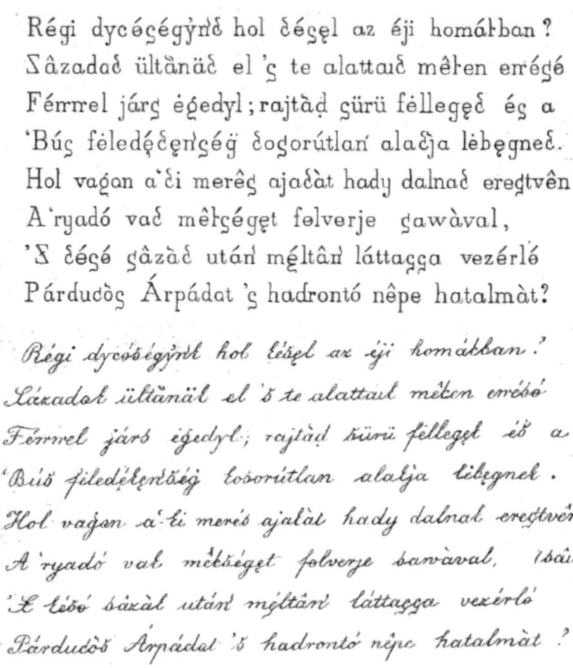

Figure 11. Examples from Dunay's second transcription system (parts of a Hungarian poem both in printed and in handwritten forms), see [7: 20-21].

The last Hungarian inventor who created an analphabetic transcription system was Gusztáv Bárczi [gustaːv baːrʦi] (his original name was Gusztáv Kratina, 1890–1964), a speech therapist who worked for the Institute of the Deaf. Later, he became an otolaryngologist and established a school for hearing-impaired and deaf children. He served also as the director of the College for Special Education. He made clear his goal in the Introduction of his book [2] which was to help his "hearing-impaired fellow human beings" when giving them a "tool" for the conscious exercise of the articulation gestures. He admitted that he had been studying speech for 17 years before started writing his book.

His book has three parts. In the first part he introduces the speech organs, muscles and their function in articulation using detailed drawings. The second and largest part contains a description of each Hungarian speech sound together with their X-ray pictures and the segment's phonetic transcription symbol in the top right corner (Figure 12). The author used white lines to divide the pictures for 4 areas the same way in all cases while Arabic numbers refer to diverse articulation organs and places of articulation to orientate deaf people.

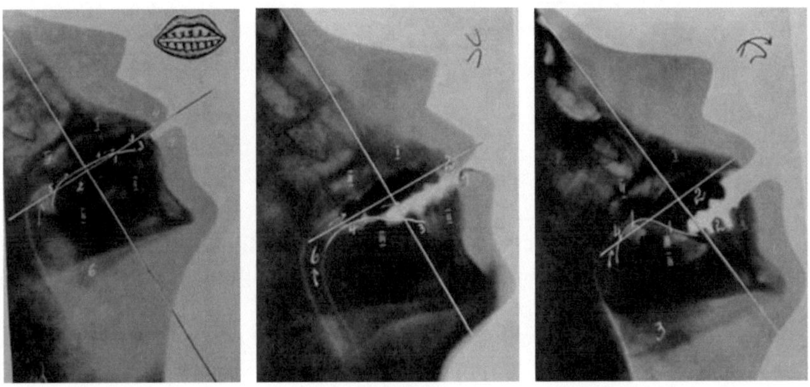

Figure 12. Characteristic positions of Hungarian speech sounds ([i-j], [f-v], [k-g]) from Bárczi's book.

The third part of the book is devoted to the author's analphabetic transcription system, explanations, and examples (Figure 13).

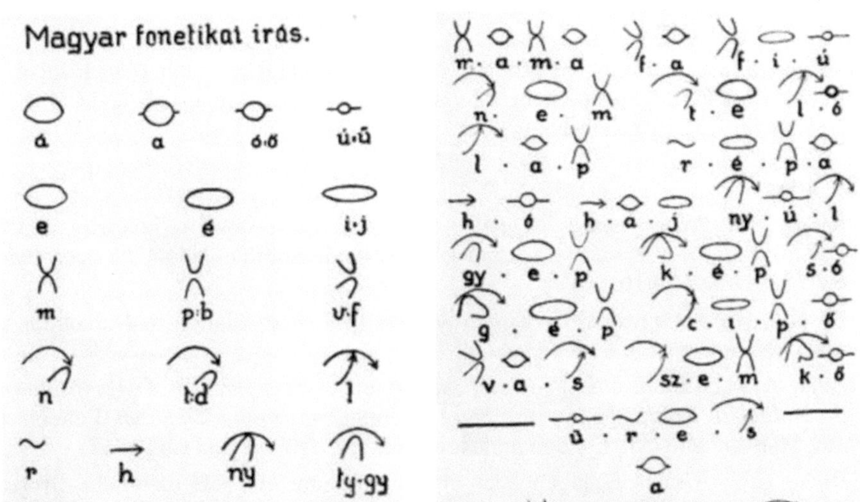

Figure 13. Bárczi's analphabetic transcription symbols (left) and transcriptions of some Hungarian words (right) (glosses: "mom, tree, boy, no, you, horse, page, carrot, snow, hair, rabbit, green, picture, salt, machine, shoe, iron, eye, stone, empty, the (definite article), and nest"), see [2: 68-69]. The title of the page with the transcription symbols is "Hungarian phonetic writing".

This system demonstrates the characteristic articulation gestures of the speech sounds. Vowels are symbolized by their lip articulation where rounded mid and high vowels are represented by the same forms of the "ellipses". Bilabial consonants are represented by the symbols of lips while dentialveolars by the symbols of teeth and lips. The palate is represented by a line with an arrow, a similar symbol represents the characteristic movement of the articulation gesture as in the case of the consonant [l]. Bárczi did not make a difference between voiced and voiceless counterparts, the phonemically short and long vowels, between [i] and [j] and between fricatives and affricates articulated at the same place.

The examples that Bárczi wrote in his book using the transcription symbols cover all Hungarian phonemes (not speech sounds!) he defined, and these appeared in monosyllables and disyllabic words. He assumed that this system would function as a kind of guide and visual reinforcement for hearing-impaired people to learn the articulation gestures of the speech sounds.

4 Conclusions

Looking back more than 100 years and stopping at some milestones of the invention of a specific phonetic writing system provides an unbelievably rich material that reflects the scientists' thoughts, knowledge on speech, creativity and motivations for their developments. These people worked usually alone without a team, practically without frequent correspondence with each other on the topic, and without the opportunity of fast access to literature, libraries, and data.

Hungarian scholars were extremely enthusiastic over the importance of their developments during 115 years from the first attempt (in 1817) to the last one (in 1928) [2], [11]. Their contribution to the international history of transcription systems is remarkable although they did not seem to influence others' inventions. Despite their relative isolation in their times [17], I am convinced that their inventions are worth remembering and they should be part of our memories on the history of phonetics.

5 Acknowledgments

This research was supported by the Ministry of Innovation and Technology of Hungary from the National Research, Development and Innovation Fund, financed under the ELTE TKP2020-IKA-06 funding scheme.

6 References

[1] ASHBY, M.: *Experimental Phonetics in Britain, 1890–1940*. University of Oxford, Oxford, 2016. https://ora.ox.ac.uk/objects/uuid:d8bbffae-8a4e-478e-ba65-0f5a5bbd66e1 [Accessed 14.10. 2019]

[2] BÁRCZI, G.: *A magyar beszédhangok képzése* [The articulation of Hungarian speech sounds]. Budapest: Franklin-Társulat, 1928.

[3] BELL, A. M.: *Visible speech: The science of universal alphabetics; or self-interpreting physiological letters, for the writing of all languages in one alphabet*. London: Simpkin, Marshall and Co., London and New York: N. Trübner and Co., 1867.

[4] BRÜCKE, E.: Untersuchungen über die Lautbildung und das natürliche System der Sprachlaute. In: *Sitzungsbrichte der kaiserlichen Akademie der Wissenschaften. Math.-naturwiss. Classe*, 3(3), pp. 181–208, Wien, 1849.

[5] BRÜCKE, E.: *Über eine neue Methode der phonetischen Transscription. Mit einer lithographischen Beilage*. Aus der K. K. Hof- und Staatsdruckerei in Commission bei Karl Gerold's Sohn, Buchhändler der Kais. Wien: Akademie der Wissenschaften, 1863.

[6] BUGÁT, P.: *Némelly éfzrevételek az újonnan tervezett helyes írás alkalmára* [Some remarks concerning the newly planned writing]. Manuscript. Budapest: Academic Library, Storage of manuscripts, 1856.

[7] DUNAY, F.: *Átiró hangrajz (Phonetica transscriptoria). Különös tekintettel a magyar nyelvhangok kifejlődésére szóban és írásban* [Transcriptional sound drawing considering the production of the Hungarian speech sounds]. Budapest: Lampel Róbert, 1901.

[8] HANSMEYER, CH.: *The History of Phonetic Alphabets* [e-book]. 2002. https://www.grin.com/document/9089

[9] JESPERSEN, O.: *The articulations of speech sounds represented by means of analphabetic symbols*. Marburg: Elwert, 1889.

[10] KEMP, J. A.: History of phonetic transcription. In: E. F. K. KOERNER & R. E. ASHER (Eds.): *Concise History of the Language Sciences*. Oxford: Elsevier Science Ltd, pp. 388–401, 1995.

[11] KYSS, A.: *Organia Sermonis [et] Scripturae universalis per Scalum et Attractionem gravitationem vocalem evidenter et mathematice determinate*. Manuscript. Budapest: Academic Library, Storage of manuscripts, Quart. Lat. 101, 1817.

[12] LAZICZIUS, G.: *Fonétika* [Phonetics]. Budapest: Királyi Magyar Egyetemi Nyomda, 1944.

[13] LEPSIUS, K. R.: *Das allgemeine linguistische Alphabet*. Berlin: W. Hertz, 1855.

[14] LUNDELL, J. A.: *Det svenska Landsmålalfabetet tillika en öfversikt af språkljudens förekomst inom svenska mål*. Stockholm: P. A. Norstedt & Söner, Kongl. Boktryckare, 1879.

[15] MACMAHON, M. K. C.: Orthography and the early history of phonetics. In: K. ALLAN (Ed.): *The Oxford Handbook of the History of Linguistics*. Oxford: Oxford University Press, 2015.

[16] MATÚSIK, A.: *Alphabetum et orthographia universalis. Ex naturae, et artis observationibus deducta ac elaborata*. Rosnaviae, 1837.

[17] ROMSICS, I.: *A Short History of Hungary*. Budapest: Osiris Kiadó, 2016.

[18] SWEET, H.: *A Handbook of Phonetics*. Oxford: Clarendon Press, 1877.

[19] TARANTELLI, V.: *Voice into Text: Case Studies in the History of Linguistic Transcription*. Ph.D. dissertation, the University of Sheffield, 2015.

[20] THAUSING, M.: *Das natürliche Lautsystem der menschlichen Sprache. Mit Bezug auf Brücke's Physiologie und Systematik der Sprachlaute*. Leipzig: Wilhelm Engelman, 1863.

[21] VÉRTES, O. A.: *A magyar leíró hangtan története az újgrammatikusokig* [The history of Hungarian phonetics up to the Neogrammarians]. Budapest: Akadémiai Kiadó, 1980.

Leaving the ivory tower: How real-life events have impacted phonetics throughout history

Angelika Braun

University of Trier, Germany
brauna@uni-trier.de

Abstract: Academic research has a reputation of being somewhat detached from reality. The metaphor of researchers being confined to the ivory tower, unable to leave it and largely unable to relate to the real world is indicative of this view. However, this view might not always be adequate. This contribution provides a number of counterexamples from the field of phonetics. Several real-world events are discussed which have had a very direct and measurable impact on phonetic research. The first example goes back to the time before phonetics was known as "phonetics". It concerns the change in attitude towards autopsies that unleashed the curiosity about human anatomy, which permitted the evolution of phonetics in the first place.

The other examples are more recent. They are often related to criminal cases in which phoneticians have acted as experts. One example is research on memory for voices, which was prompted by the kidnapping of the Lindbergh baby in 1932. A further example concerns the development of the spectrograph, which was part of the American WW II wartime strategy. The next major event affecting phonetic research was the impeachment trial of Richard M. Nixon, in the course of which a clandestine recording made in the White House had to be authenticated. In 1989, the supertanker *Exxon Valdez* grounded in Prince William Sound, Alaska. In this case as well as a similar one in Germany, the question of whether the captain was drunk at the time of the incident played a major role in the court cases. Finally, the various attempts at establishing a German pronunciation standard are reviewed in light of political developments in recent history.

1 Introduction

When academics try to explain to ordinary people what they are doing, they tend to meet with limited understanding. It is not easy to communicate why for instance proving a certain mathematical formula or studying deep-sea marine life should be allotted large grants. Compared to these topics, the phonetic sciences have always had closer ties to the real world. Early examples are the teaching of the deaf, the teaching of pronunciation to L2 learners or, more recently, the fitting of cochlear implants. In some instances, the approach has turned the default case of research planning upside down: instead of generating a theory and testing it against observation [58: 390], there can be no doubt that a whole wave of phonetic research on particular topics was prompted by real-world events. Some selected examples will be covered in this contribution. The aim of this paper is thus to demonstrate that sometimes new steps in phonetic research were brought about by specific real-world events as opposed to being theory-driven.

2 An early example – autopsies as a prerequisite for phonetic research

2.1 Some facts

Some very early developments in phonetics were influenced by historical events. Strictly speaking, what we now consider the beginnings of phonetics in the 16th century would not have been possible without a change of mind with regard to autopsies. In third-century B.C. Egypt, autopsies, sometimes even vivisections, were used to gain insight into human anatomy. Herophilos of Chalcedon (335–280 B.C.) and Erasistratos of Ceos (310–250 B.C.) were the most famous representatives of that tradition [17: 93]. Unfortunately, their knowledge was lost to future generations in a large fire which destroyed the famous library of Alexandria in 47 B.C. [2: 4].

Claudius Galenus (129–199), who is considered one of the major anatomists in European antiquity, gained much of his knowledge about human anatomy in his capacity as "team physician" of the Roman gladiators and by dissecting animals. However, he did not venture to touch human corpses [2: 5]. In the Middle Ages, all this knowledge was lost. Autopsies were considered unethical, and especially the Church was strictly opposed to them. One of the most outspoken representatives of the Church was Augustinus (354–430) who harshly condemned autopsies. A ruling by Pope Bonifaz VIII in 1299 was interpreted along those same lines [17: 93-94].

The Renaissance brought a change in attitude towards autopsies. They became popular at Italian universities in particular. While they were carried out in secrecy at first at the University of Salerno, they soon developed into *spectacula* where chairs were set up in a manner reminiscent of a theater.[1] While a "dissector", often a barber, did the actual opening up of the body, the professor was seated in an elevated chair – removed from the smell – and recited Galenus' works. An assistant was supposed to point to the structures which were mentioned by the professor [17: 94].

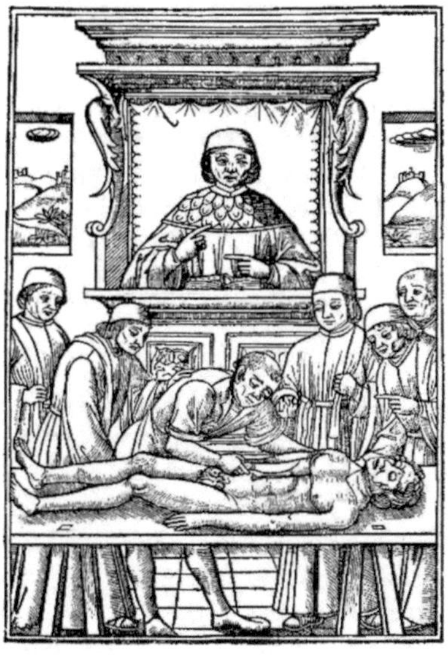

Figure 1. Typical scenario of an autopsy in Renaissance times: Mondinus (1275–1326), a well-known anatomist of his time, teaching anatomy. Wood carving in "Anathomia Mundini", Geneva 1519 (reprinted with permission from [57: 425]).

[1] Please note that, till today, the operating room is often called the "operating theater".

Tickets for these shows were extremely popular even though the narrations were usually unrelated to what was actually happening on the dissection table. This "hoax" was exposed in 1535 by a young student in Paris by the name of Andreas Vesalius (1514–1564), who ventured to step up against the famous anatomist Jacobus Sylvius[2]. Vesalius is considered to be the father of modern anatomy. He was born in Flanders and became an outspoken critic of Galenus, rather because he realized that the latter had not gained his knowledge from dissecting humans but animals instead. He received a professorship in Padova at the age of 23 and was very successful, because he did both the dissecting and the narrating, with these two corresponding for the first time ([2: 9], [17: 94-95]).

2.2 How phonetics comes into play

Vesalius is considered the founder of scientific anatomical studies [2: 9] and is thus at the same time a pioneer in phonetics in the sense that he expanded the knowledge about the anatomy of the organs involved in speaking and listening. He overcame the Galenic tradition and relied on data gained from dissecting human bodies. This brought about essential knowledge gain in the anatomical domain. The physiology of speaking and hearing, however, remained largely in the dark – the exact mechanism of vocal fold vibration, let alone the physiology of hearing, was not described until well into the 20th century [53; 55; 59].

2.3 Conclusion

Anatomical studies are a prerequisite for phonetic research. Autopsies have always been the prime source of knowledge about human anatomy including the vocal organs. Not even the development of MRI has been able to replace the immediate inspection of the structures involved. Thus, the findings of the early anatomists were indispensable for any phonetic description of sound production and perception.

3 The kidnapping of the Lindbergh baby

3.1 Some facts

On 1 March 1932, the 20-month-old son of Charles and Anne Morrow Lindbergh disappeared from his bed. Charles Lindbergh was a world-famous aviator known for having crossed the Atlantic Ocean solo in a small passenger plane in 1927. In the boy's room, a note was found which was written in poor English and demanded $ 50,000 in ransom. In the course of the next five weeks, the parents received another twelve ransom notes. Some sources also mention phone calls made by the kidnapper and answered by Charles Lindbergh [58]. On 2 April 1932, ransom money was handed over by John F. Condon, who acted as an intermediary for the Lindberghs. On that occasion, Charles Lindbergh had a chance to listen to the kidnapper's voice from a distance. However, the whole transaction turned out to be a hoax. The child was not found where the man who received the ransom said he was. Instead, the corpse of Charles Augustus Lindbergh, Jr. was found close to the Lindbergh home on 12 May 1932. It turned out that he had been killed shortly after the kidnapping [61]. Bruno Richard Hauptmann, a carpenter with German roots, was charged with the kidnapping and murder of the child. The evidence against him was mainly circumstantial. Part of it consisted of Charles Lindbergh's testimony on the kidnapper's voice [46]. Hauptmann denied any involvement in the case, and yet he was

[2] The Sylvian fissure, which separates the frontal from the temporal lobe in the brain, was named after Sylvius.

convicted and – after several unsuccessful attempts to appeal the verdict – eventually electrocuted on 3 April 1936.

3.2 How phonetics comes into play

The conviction of Bruno Hauptmann rested, to a large extent, on the testimony by Charles Lindbergh. He testified to having recognized Hauptmann's voice as that of the man whom he had heard on the occasion of the money handover when he was given a chance to listen to him shortly before taking the stand. His identification was based on nine syllables of speech ("Hey Doctor, over here, over here") with a time lapse of close to three years between familiarization and recognition. Still, the jury went along with his statement and handed down a guilty verdict. The defense questioned the recognition procedure. It was pointed out that there was no scientific evidence to the effect that recognition was possible under the given circumstances. Prompted by this case, the psychologist Frances McGehee of Johns Hopkins University, Baltimore, MD carried out pioneering work on memory for voices [38], [39], [56]. She conducted a series of experiments some of which simulated the setting of the Lindbergh case. In doing so, she addressed most of the principal questions which are still at issue today when a so-called voice line-up is to be carried out: the a priori possibility of choosing a number between one and five (position effect), the effective reduction of the line-up size by including a voice which is very dissimilar to the others, the issue of memory type (incidental vs. intentional) and presentation format (live or taped), the effect of the number of voices present in the line-up, the effect of a target-absent scenario on the recognition rate, the effect of voice disguise, the role of race and gender in recognition, the performance of musically and linguistically trained listeners, and, last but not least, the time lapse.

McGehee found that recognition accuracy deteriorated from a recognition rate of 83% one day after familiarization with the voice to a mere 13% five months later. She also found that recognition rates go down if the number of voices in the line-up is increased. While disguise presented a problem to listeners, a foreign accent did not. Men generally outperformed women at recognition tasks [38].

Her research has been criticized for reverting in part to a within-subjects design by assigning some listeners to more than one group [58: 391]. However, even though the research paradigm may not comply with today's standards in all respects, she has to be given credit for identifying the relevant issues and for addressing them in a large number of experiments.

3.3 Conclusion

To this day, there is hardly any publication on the subject of speaker identification by lay listeners that does not pay tribute to McGehee's pioneering work and the ties to the Lindbergh case (cf. e.g. Clifford [13: 375]; Deffenbacher et al. [15] and especially Yarmey et al. [58]). To this date, the issue of victims and witnesses recognizing a perpetrator's voice remains largely unresolved. The existing studies are largely "academic", and there is a paucity of projects with a clearly defined forensic perspective. Specifically, it would be highly desirable to develop an independent test which could provide a measure of a given witness's general ability to recognize voices. That would solve the principal question behind any line-up in which an identification is made – to what extent can the witness be relied on? In the meantime, the issues raised by McGehee in the 1930s should be addressed and re-investigated in terms of time lapse, memory type and voice characteristics in particular.

4 The Sound Spectrograph

4.1 Some facts

During World War II, the American military worked closely together with the nation's leading think tanks, among them Bell Labs at Mahwah, N.J. After Japan entered the war with the attack on Pearl Harbor on 7 December 1941, Japanese and German submarines posed a continuous threat to the US navy. It was of prime strategic importance to be able to trace the movements of the submarines in order to know what the enemy forces were up to. There was nothing to go by except the radio communication between the vessels, which was intercepted by the US forces on a regular basis.

4.2 How phonetics comes into play

Until the 1930s, the kymograph and the cathode ray oscilloscope were the main instruments that were available for visualizing different aspects of speech behavior. This changed in the early 1940s, when Ralph K. Potter developed the Sound Spectrograph. According to Koenig [31: 1], a prototype was finished in the spring of 1941. Potter filed the patent application on 14 April 1942. In the course of testing the new device it must have occurred to Potter that the spectrograms did not only show properties of different sounds, but they could also be used to depict reflections of different vocal tracts, i.e., differences between speakers. This is hinted at, but never elaborated in various post-war papers by Potter and his colleagues (cf. Potter [43], [44], and Kopp & Green [34]). The spectrograph gained strategic importance and was developed as a secret device which could be used to produce "voice prints" of the German and Japanese radio operators and thus to identify them (cf. Braun [7], [8]). A task force was formed that developed a framework for implementing the spectrograph as a speaker identification device. Figure 1 shows an excerpt from a secret memorandum entitled "Field Uses of Voice Identification" of 31 May 1944 which was authored by Ralph Potter. It evidently was part of a preliminary version of the final report submitted to the military on 18 August 1944 [21].

The voices of individuals differ. These differences in some degree are familiar to the ear and can be recorded and classified by the sound spectrograph. Such records and classifications may be used as a basis for voice identification which, with radio direction finding methods, would make it possible to follow movements of enemy personnel using radio telephone facilities and hence to trace changes in disposition of enemy units. The use of voice for this purpose is analogous to that which has already been developed for manual telegraph.

Figure 2. Facsimile from a memo of 31 May 1944, prepared by Ralph Potter [42: 1].

In the final report that passage reads as follows (cf. Figure 3):

By detecting the characteristic differences in the way individual radio telegraph operators send telegraph signals and correlating these with direction finding data, it is possible to follow the movements of these operators from place to place and therefore determine indirectly the movements of enemy groups with which they are associated. Similarly, by identifying the speech of individuals talking over radio telephone circuits, it should be possible to associate these identifications with direction finding data and to determine the movements of enemy troops or other mobile groups.

Figure 3. Facsimile from Gray & Kopp ([21, p. 1][3]

This application of the sound spectrograph was not published until well after the war. In the meantime, the spectrograph was said to have been developed to enable the deaf to communicate over the telephone. This is completely unrealistic but was used as a cover for many years and even continued in the famous monograph by Potter, Kopp and Green [45].

4.3 Conclusion

In times of war, the ties between military goals and science were particularly close. This sped up the development of the spectrograph, and it prompted research into the potential of spectrograms displaying speaker specificity.

It seems that the end of World War II put a preliminary halt to the spectrograph as a tool in voice comparison. After all, the relevant documents were still classified. Soon after their declassification in 1960, Lawrence Kersta, an engineer who had also worked at Bell Labs, published the "voiceprint" method of speaker identification, thereby misrepresenting Potter and his colleagues [8], [29], [30]. This, in turn, brought about more studies on the "voiceprint" method of forensic voice comparison (cf. e.g. Braun [7]). To this day, the software tools which are based on the sound spectrograph are still the principal means of visualizing and analyzing human speech.

5 The Watergate Affair

5.1 Some Facts

On 17 June 1972, five men were caught breaking into the Democratic Party headquarters in a building complex called Watergate in Washington, D.C.. It turned out that they had not only stolen sensitive documents but also planted surveillance devices in order to assure the re-election of President Richard Milhous Nixon, who was a Republican. The question then was if and when the President had learned about this criminal offense which had been masterminded by the so-called Committee to Re-Elect the President (C.R.P.). The investigation was led by John Sirica, then District Judge in Washington, D.C.. Congress was also looking into the affair: On 25 May 1973, the Senate Select Committee appointed a Special Prosecutor by the name of Archibald Cox. On 16 July 1973, the former Presidential Appointments Secretary Alexander Butterfield testified that President Nixon had had all conversations taped that had taken place in the Oval Office since 1970, including those conducted over the telephone [1: 857]. The controversy revolved around one conversation between the President and his then Chief of Staff, John Haldeman, on 20 June 1972. On 11 April 1974, the House Judiciary Committee subpoenaed 42 tapes, but Nixon refused to supply them, invoking Executive Privilege [1: 876]. In what

[3] Note the reserve that is contained in the wording "should be possible" for speaker identification as compared to "is possible" with respect to identifying telegraph signals.

has come to be known as the Saturday night massacre, President Nixon fired his Attorney General Richardson for his refusal to fire Cox and had his successor fire Cox instead, who was replaced by Leon Jaworski [1: 866]. Judge Sirica then subpoenaed 64 tapes, and once again the White House did not comply. The issue of the tapes went all the way to the Supreme Court, which rejected Nixon's reasoning by an 8:0 vote on 24 July 1974.

Eventually, the tapes, including the one of interest here, were served by the White House. On one tape, there was an 18 ½ minute gap which contained no speech but only low-frequency hum. The refusal to provide the tapes had in the meantime prompted an impeachment trial against Nixon. It was formally initiated on 6 February 1974 [1: 873]. Richard Milhous Nixon did not wait for the outcome. He resigned on 9 August 1974.

5.2 How phonetics comes into play

The million-dollar question in conjunction with the mysterious 18 ½ minute gap in the recording of 20 June 1972 was, of course, whether it was accidental or the result of a volitional attempt to obstruct justice. The President's personal secretary, Ms. Rose Mary Woods, offered an explanation: On 27 November 1973, she testified that she might have inadvertently erased about five of the 18 ½ minutes in question when she was trying to transcribe the tape [1: 868].

Judge John Sirica proceeded to appoint a panel of six senior figures in audio engineering, physics, and also phonetics, who were to analyze the ominous 18 ½ minutes. They were Richard H. Bolt of Bolt, Beranek, and Newman, Inc., a former professor of Acoustics at M.I.T., James Flanagan, Head of the Acoustics Research Department at Bell Labs, John McKnight, a private consultant, formerly with Ampex Corporation, Thomas G. Stockham of the University of Utah, formerly at M.I.T., Mark R. Weiss, Vice President for Acoustics Research at the Federal Scientific Corporation (F.S.C.), and Franklin Cooper of Haskins Labs (cf. [3: C.1-C.7]). This panel started working on 30 November 1973 and produced a final report on 31 May 1974.[4] The reason these senior figures in acoustics, phonetics, and audio engineering were assembled is that this evidently was the first time that an investigation of this nature was carried out in the US, and probably worldwide. Never before had this type of question arisen, at least not in such a high-profile case. The experts were supplied with a total of nine tape recorders (seven SONY 800B which had been used to record the conversations in the Oval Office, one UHER 5000 with a foot pedal[5] from Ms. Woods' office, and one UHER 5000 which was in use by the Secret Service) and the famous tape. The following questions were addressed in the investigation:

- Is the recording in question an original or an – edited – copy which contains erasures or splices?
- Where does the buzzing noise in that 18 ½ minute section originate from? Was it recorded continuously or in several chunks?
- Can it be determined whether there was speech "underneath" the buzzing sound? If so, can the speech be recovered? (cf. [3: 1]).

The authors of the report describe in detail the methods they applied (cf. [3: 7-20]). They range from critical listening to acoustic analyses of the on- and off-transients of the record head and

[4] Incidentally, all their work was carried out under the supervision of US Marshalls, who never left the evidence unattended (cf. [3: B.1]).

[5] The foot pedal is used to start and stop the machine when transcribing so that the transcriber does not have to take his/her hands off the typewriter. It cannot be used, however, to erase and record over.

the erase head of the tape recorders in question and "magnetic development", i.e., the application of ferrofluid to the tape in order to render the magnetic imprints visible. These methodological elements have since become standard procedure for the authentication of analogue audio tapes (cf. [29], [30], [32]).

The panel stated its conclusions unanimously and with no lack of clarity. They concluded that the tape was an original which had, however, been tampered with. This conclusion was primarily based on acoustic phonetic analysis and the inspection of magnetic patterns on the tape. Between five and nine erasures had been willfully executed.

The buzzing noise was determined to originate from a power line leakage. This finding rested on precise frequency analysis. It was established that there must have been speech signal before the erasures, but the experts were not able to recover it. This could be concluded from an analysis of the short snippets which were located in the 55ms gap between the erase head off and erase head on marks (cf. [3: 15-16]). Short as they were, the acoustic structure within those snippets clearly showed that language had been contained on those parts of the tape before they were erased.

The experts also checked the possibility of accidental erasure, and they came to the conclusion that this was basically impossible because "manual operation of a keyboard control is required to put the recorder into erase-record mode" [3: 44]. On 18 February 1974, the Nixon camp supplied the panel with an unsolicited report by Allan D. Bell, president of Dektor, a small company specializing in surveillance equipment [47]. That report stated that it would have been possible for Ms. Woods to erase part of the tape accidentally by stretching to pick up the phone while her foot remained on the tape recorder foot pedal. The panel of experts refuted that, and it was not followed up in court.[6]

5.3 Conclusion

Authenticity analyses have since become an integral part of forensic phonetic work (cf. [5], [10], [20], [24], [32], [33], [37]). Most publications covering this aspect point to the Watergate affair as the starting point of forensic authenticity examination. The fact that this was the first time that the question of authenticity arose in a high-profile case has helped to establish the methodology used by, e.g., Koenig [32] and Boss et al. [4] in later cases.

In the meantime, methods have changed due to digitization [33]. Therefore, this is now history, unless an old case resurfaces, as just recently happened in Germany, and the reports from the 1990s are re-examined [60].

6 About drunk drivers and drunk seamen

Until the 1980s, studies on the effect of alcohol on speech were scarce. The bulk of research was medical by nature. Based on a thorough literature search, Chin and Pisoni [12] count a total of 22 papers on alcohol and speech between 1915 and 1982 (p. 167). There were two main reasons why that changed dramatically in the 1980s.

6.1 Drunk drivers

In the early 1980s, the US automobile industry was trying to install a safeguard against driving while intoxicated into its vehicles. They were hoping to develop a device which would determine whether or not the driver was sober, based on a short sequence of speech. If the algorithm

[6] Decades later, in 1995, a memorandum was found in Ms. Woods' papers which confirmed that the erasure was volitional [26].

detected effects of alcohol, the engine would not start. As is quite evident to any phonetician, such a device is not a realistic option for a number of reasons, among them that there is no simple one-to-one relationship between speech behavior and the (degree of) intoxication (see below). In fact, it was never implemented. However, General Motors Research Laboratories had commissioned researchers at Indiana University in Bloomington to determine the possibility of detecting whether a potential driver was intoxicated and, as a consequence, putting a lock on the ignition [12: 168]. This resulted in a certain expertise by those researchers on the subject of alcohol and speech when the next incident happened [40].

6.2 A drunken captain - the grounding of the *Exxon Valdez*

6.2.1 Some facts

On 24 March 1989, an American oil tanker, the *Exxon Valdez*, grounded on Bligh Reef in Prince William Sound, Alaska. Some 10,836,000 US gallons of crude oil leaked into the environment [12: 285]. About 2,100 kilometers of coastline were polluted; hundreds of thousands of birds were killed. It is considered the worst oil spill worldwide to date. The damage amounted to billions of dollars. The ensuing investigation revealed that the captain, Joseph Hazelwood, had not been on the bridge during the critical hours. Instead, he had withdrawn to his quarters, leaving the vessel to the third mate Gregory Cousins to maneuver through the difficult territory. Further, it turned out that Captain Hazelwood had a history of alcohol abuse. He had lost his driver's license because of repeated DUI ("driving under the influence") offences and admitted to having had a couple of drinks earlier that evening. It wasn't until nearly 10 hours after the incident that blood and urine samples could be taken from the captain. Hazelwood, it turned out, had a blood alcohol concentration (BAC) of .061% at that point in time. Witnesses testified that his breath had smelled of alcohol that night, but they also said that they had not observed any signs of intoxication in his behavior although his eyes had been watery [12: 286].

Joseph Hazelwood was fired by Exxon and charged with three counts of misdemeanor (reckless endangerment, operating a vessel while intoxicated, and negligent discharge of oil) and one felony (criminal mischief; cf. [12: 292-293]). He denied all charges and went to trial in February1990. In the criminal trial, he was acquitted of all charges but negligent discharge of oil, but he was also facing a civil trial because local fishermen and property owners had filed suit against him personally as well as against Exxon, asking for damages.

Incidentally, a similar accident had taken place in Germany three years prior to the Exxon Valdez disaster: on 3 January 1986, the British MS *Waylink* collided with the Panamanian tanker *Brady Maria* on the Elbe River. About 260 tons of residual fuel oil were spilled into the water and caused major damage to the marine wildlife. It turned out that the pilot on the *Waylink,* who was known for his alcohol addiction, was completely drunk when tested shortly after the incident (BAC > .2%). Speech samples from the pilot around the time of the collision were available for analysis and prompted a research project [36].

6.2.2 How phonetics comes into play

In the course of the *Exxon Valdez* investigation, it occurred to the authorities that there were a number of recordings of Captain Hazelwood's speech on marine radio communication tapes at various points in time: He had talked to the Coast Guard 33 hours and then again one hour prior to the grounding, a few minutes after the incident, and one hour as well as nine hours after the grounding. The question arose whether it was possible to establish whether he was inebriated on any of these occasions based on these samples of his speech, and if the level of intoxication could be determined. The National Transportation Safety Board (NTSB) investigated the case. Three groups of experts were consulted: The NTSB's own audio experts Brenner and Cash [9], Mark B and Linda C. Sobell of the Addiction Research Foundation in Toronto, Canada, and a group of researchers from the Speech Research Laboratory at Indiana University ([27], [41], [49]).

The first and the last group of experts published their results. They concluded that "the speech changes exhibited by the master of the Exxon Valdez were consistent with changes observed when a talker is impaired by alcohol" [12: 230].

A controversy arose about the admissibility of this type of evidence under US law. Tanford et al. [49] argued in favor, stating "that the scientific evidence of intoxication in the *Exxon Valdez* case was sound and reliable" [12: 297], whereas Hollien [25][7] concluded "that the speech and voice data did not support the findings of the NTSB and its consultants that Hazelwood was intoxicated around the time of the grounding" [12: 300].

That same question of whether (the degree of) inebriation could be established from speech samples had come up in a German court two years earlier in conjunction with the case mentioned above. However, there was not much previous research that the German experts in this case could rely on ([25], [27], [36]). Both these cases prompted a major research project by voice experts from the Federal Criminal Police Office (BKA) on the effects of alcohol on human speech.

In the German case, the pilot's license was revoked for five years whereas Hazelwood's was suspended for nine months only. However, he was sentenced to $ 50,000 in damages and to 1,000 hours of community service [12].

6.3 Conclusion

Even though the voice evidence did not play a major role in either the criminal or the civil trial, mainly because the prosecution stated that it was made aware of the findings too late [12: 292]. Chin and Pisoni [12] summarize that "the audiotape recordings of radio transmissions between Captain Hazelwood and the Coast Guard Vessel Traffic Center were the sole objective physical evidence from the time around that of the grounding" (p. 312). Even though the voice evidence did not decide the case, it did set off a series of scientific experiments which have put researchers in a position to argue more reliably in court. After having carried out extensive research in this domain, the BKA group concluded that speech is a sensitive indicator of intoxication. In fact, it proved more sensitive than some of the common neurological tests such as walking a straight line. On the other hand, results show clearly that speech behavior allows no estimate of the degree of intoxication due to confounding factors such as individual tolerance, food intake or fatigue. If all of these are known, it would be conceivable to replicate the circumstances, though. While there may not be an easy solution to the basic question, the *Exxon Valdez* case "does represent the movement of basic science into a fairly well-known real-world situation" [12: 312].

7 The standardization of German pronunciation

7.1 Some historical facts

It is virtually impossible to understand the process of standardizing German pronunciation without looking at the political developments in that country. As opposed to countries such as England and France which had been political entities for centuries, the area that Germany now covers was split up into literally dozens of kingdoms, principalities etc. It was Bismarck who managed to unite Germany and form an empire following the Franco-Prussian war in 1871.

[7] As is customary in adversarial systems like for instance in the US, there were opposing experts, in this case from the University of Florida.

This empire broke up as a consequence of WW I. After WW II, Germany was divided up among the allied powers, and in 1949 the groundwork was laid out for the – western – Federal Republic of Germany (FRG) and the – eastern – German Democratic Republic (GDR) as independent states. This historical rift ended with German reunification in 1989, when the GDR effectively ceased to exist and joined the FRG.

7.2 How Germany ended up with five pronouncing dictionaries

Establishing a pronunciation standard for German has proven to be a tedious and highly politicized endeavor. It is worth taking a moment to consider how other countries went about defining their standard variety. In centralist France, the Académie Française was quick to determine that the pronunciation by the educated speakers in the greater Paris area, was to be the standard [11]. The Comédie-Française served as the model for stage pronunciation. In England, the socially but not regionally marked pronunciation in the expensive private boarding schools and colleges was accepted as Received Pronunciation in the 19th century. Until fairly recently, it was totally inconceivable to hear anything but RP on, say, the BBC.[8] In Germany, the situation was different, mainly because there had been no Germany until 1871. The newly united country needed to shape its identity, and one of the measures to that effect was to standardize orthography and pronunciation ([19]; [35: 8]). Those in charge went about it in a very "German" manner – i.e. a task force was formed that was to establish the "proper" pronunciation. That group consisted of six representatives from academia and theater, among them Theodor Siebs, who was also responsible for the publication of the dictionary. Most of them came from the north of Germany, and that group decided to use stage talk as a model. To cite an example, this meant that it was mandatory to pronounce the /r/ as an apical trill in all positions, i.e. also postvocalically [48][9]. There was also a strong bias toward "northern" characteristics, although no clearly defined region served as a model. A phonetician by the name of Wilhelm Viëtor opposed this and other prescriptive elements and published a competing pronouncing dictionary in 1890 based on a survey among educated speakers [54]. This publication never achieved a relevant degree of public attention, not least because *the Siebs,* as it had come to be known by that time, was recommended for use in schools. It appeared essentially unchanged in 18 editions, and its role as the "standard" throughout the first half of the 20th century was strengthened. Only the 19th edition [16], besides removing the term "stage pronunciation" ("Bühnenaussprache") from the title, adopted a "moderated standard pronunciation" ("gemäßigte Hochlautung") as an acceptable alternative to the stage norm.

In 1953, a meeting of delegations from both sides of the Iron Curtain took place in Frankfurt with the aim of developing a joint strategy for revising the outdated *Siebs*, but the delegations could not agree on a common set of guidelines to be followed. This is where politics comes in – the Western researchers insisted on a largely prescriptive process whereas the East German phoneticians from the University of Halle pursued a usage-based, descriptive approach which was more in line with the "democratic" self-perception of communist Germany. This split was now emphasized by the East German side because it regarded the emerging language gap between the two German states as a necessary consequence of the two ideologies on which they were based ([14], p. 120). This development increased the number of pronouncing dictionaries by two – one from Halle [22], and another authored by phoneticians dissenting from the firm

[8] Various authors have pointed out that "RP" is not quite as uniform as the term suggests. Wells [56] describes no less than five sub-varieties of RP: Conservative RP or U-RP (as spoken by the "real" aristocrats and members of the Royal Family), mainstream RP (as then spoken in the BBC), advanced RP (as spoken by the younger London upper class), near-RP (as spoken by the ambitious middle class) and adoptive RP (as spoken by those aspiring to rise socially and trying a little bit too hard to conform to RP rules).

[9] This single pronunciation feature may have contributed to Hitler's speeches sounding so unpleasant irrespective of the content.

position held by the editors of the *Siebs,* Helmut de Boor and Paul Diels, in West Germany. The latter group was headed by Max Mangold from Saarland University, who – like the East German researchers – tried to introduce descriptive elements. The resulting dictionary was published as part of the DUDEN book series [18].

Following German reunification in 1989, the issue came up again. Finally, the West German delegates were ready for a usage-based approach. A monstrous pronouncing dictionary project was initiated which essentially was to follow the principles developed in Halle, but it was at the same time to include several stylistic levels and was to be accompanied by sound files representing different speaking situations of varying formality for each entry. Halle was in charge of the codification; the University of Cologne was responsible for producing the sound files. Briefly, the project proved to be much too big to be feasible. The publication that emerged was the *Deutsches Aussprachewörterbuch* [35], published by speech scientists from Halle. It turned out to be a much less comprehensive product than projected, which undoubtedly has its merits but has also met with criticism.[10]

7.3 Conclusion

The codification of German pronunciation reflects some aspects of German history in a nutshell. This extends from the 19th century aspirations to unify language usage to the ramifications of the competing political systems in East and West Germany up to the developments that came with reunification. The different approaches to standardizing pronunciation in the two German states represent the respective political philosophy, and it is indicative of a rift within the West German society that there were two competing pronouncing dictionaries in the FRG, both publishing new editions in parallel.

8 Final remarks – phonetics and real life

Clearly, phonetics is not confined to the ivory tower. The main reason for this is the nature of the object of study, i.e., all aspects of spoken language including paralinguistic events. Since communicating is part of everyday life, the study of communication touches upon people's everyday lives. In this contribution, it has been demonstrated how real-life events have fostered and sometimes even initiated phonetic research.

On the other hand, though, there are also examples of researchers turning to everyday events as objects of their study. One of the prime examples is Jürgen Trouvain. He has had the courage to address the phonetics of sports commentaries, laughter, and clicks [50], [51], [52]. Another example is the analysis of the Queen's Christmas messages [23]. Research on these topics tends to be regarded as entertaining rather than scientific. Therefore, it has not always received the recognition it deserves from the scientific community because these topics may not be perceived as being at the center of interest to phoneticians.[11] This contribution is intended to expressly encourage this kind of enterprise. It is this very ability to relate to everyday phenomena which will keep rare subjects such as phonetics alive and which will allow it to reach out to people beyond academia.

---

[10] For instance, the most common pronunciation of the prevocalic and intervocalic /r/ is termed "velar" in the *Deutsches Aussprachewörterbuch* [35], which seems hardly tenable in view of the fact that the trill (and also the fricative) produced in the back of the oral cavity are clearly uvular.

[11] Many years ago, this author studied grunting in tennis from a phonetics perspective [6], which has met with much more interest in the medical community than in the phonetics community.

Acknowledgements

A warm word of thanks goes to Katharina Zahner for valuable comments on a draft version of the paper, and to Anna-Maria Duplang for proofreading it and checking the English.

References

[1] AMSTER, L.: *Chronology of Watergate-related events*. In: The White House Transcripts. The full text of the Submission of Recorded Presidential Conversations to the Committee on the Judiciary of the House of Representatives by President Richard Nixon. With an introduction by R.W. APPLE of The New York Times. Toronto: Bantam, pp. 813–877, 1974.

[2] BECKER, K.W.: *Anmerkungen zur Geschichte der anatomischen Sektion*. Text zum Katalog der Ausstellung KunstOrt Anatomie. Künstler auf Visite. 23. Mai bis 21. Juni 2002 im Anatomischen Institut der Universität des Saarlandes, 11pp., 2002.

[3] BOLT, R.H., COOPER, F.S., FLANAGAN, J.L., MCNIGHT, J.G., STOCKHAM, T.G. & WHITE, M.R.: The EOB tape of 20 June, 1972. Report on a technical investigation conducted for the U.S. District Court for the District of Columbia by the Advisory Panel on White House tapes, 51pp., 1974.

[4] BOSS, D., GFROERER, S. & NEOUSTROEV, N.: A new tool for the visualization of magnetic features on audiotapes. *Forensic Linguistics*, 10, pp. 255–276, 2003.

[5] BRAUN, A.: Die forensische Analyse von Stimme und Sprache. In: H. GUNDERMANN (Ed.): *Die Ausdruckswelt der Stimme. Erste Stuttgarter Stimmtage*. Heidelberg: Hüthig, pp. 88–102, 1998.

[6] BRAUN, A.: Phonetische Betrachtungen zu einem Phänomen im Tennissport: Eine explorative Studie zum *grunting*. In: C. MAUELSHAGEN & J. SEIFERT (Eds.): *Sprache und Text in Theorie und Empirie. Beiträge zur germanistischen Sprachwissenschaft.* Stuttgart: Steiner, pp. 198–208, 2001.

[7] BRAUN, A: From Visible Speech to Voiceprints – Blessing and Curse of Sound Spectrography. In: M. VAINIO, J. SIMKO & R. AULANKO (Eds.): *HSCR 2017. Proceedings of the Second International Workshop on the History of Speech Communication Research.* Helsinki, August 18–19, 2017. Dresden: TUDpress, pp. 68–78, 2017.

[8] BRAUN, A.: From Visible Speech to Voiceprints – The Missing Link: In: M. PUCHER, J. TROUVAIN & C. LOZO (Eds.): *HSCR 2019. Proceedings of the Third International Workshop on the History of Speech Communication Research.* Vienna, September 13–14, 2019. Dresden: TUDpress, pp.123–136, 2019.

[9] BRENNER, M. & CASH, J.R.: Speech analysis as an index of alcohol intoxication – The Exxon Valdez accident. *Aviation, Space, and Environmental Medicine*, 62, pp. 893–898, 1991.

[10] BROEDERS, A.P.A.: Forensic speech and audio analysis. Forensic Linguistics 1998-2001. In: *Proceedings of the 13th INTERPOL Forensic Science Symposium, Lyon, France, Vol. 2*, pp. 54–84, 2001.

[11] BUFE, W.: Aussprachenormen im Gegenwartsfranzösischen am Beispiel der chroniques de langage. Diss. Saarbrücken, 2014.

[12] CHIN, S.B. & PISONI, D.B.: *Alcohol and Speech*. San Diego: Academic Press, 1997.

[13] CLIFFORD, B.R.: Voice identification by human listeners. On earwitness reliability. *Law and Human Behavior*, 4(4), pp. 373–394, 1980.

[14] CLYNE, M.: German as a pluricentric language. In: M. CLYNE (Ed.): *Pluricentric languages. Differing norms in different nations*. Berlin/New York: de Gruyter, pp. 117–148, 1991.

[15] DEFFENBACHER, K. A., CROSS, J. F., HANDKINS, R. E., CHANCE, J. E., GOLDSTEIN, A. G., HAMMERSLEY, R. & READ, J. D.: Relevance of voice identification research to criteria for evaluating reliability of an identification. *The Journal of psychology,* 123(2), pp. 109–119, 1989.

[16] DE BOOR, H., MOSER, H. & WINKLER, C. (Eds.): *Siebs. Deutsche Aussprache. Reine und gemäßigte Hochlautung mit Aussprachewörterbuch.* 19., umgearbeitete Aufl. Berlin, 1969.

[17] DIALLO-DANEBROCK, R., ABBAS, M., GROß, D. & KELLNER, U.: Geschichte der anatomischen und klinischen Obduktion. *Der Pathologe*, 40(1), pp. 93–100, 2019.

[18] DUDEN. *Aussprachewörterbuch. Wörterbuch der deutschen Standardaussprache*. Bearb. von Max MANGOLD und der Dudenredaktion unter Leitung von Paul Grebe. Mannheim, 1962.

[19] DURRELL, M.: Standardsprache in England und Deutschland. *Zeitschrift für Germanistische Linguistik*, 27(3), pp. 285–308, 1992.

[20] FRENCH, J.P.: Authenticity examinations of audio recordings. *Proceedings of the Institute of Acoustics*, 1991.

[21] GRAY, C.H. & KOPP, G.A.: Voice Print Identification. In: *Bell Telephone Laboratories Report*, pp. 1–14, 1944.

[22] *Grosses Wörterbuch der deutschen Aussprache.* Hg. Von dem Kollektiv E.-M. KRECH, E. KURKA, H. STELZIG, E. STOCK, U. STÖTZER & R. TESKE. Leipzig, 1982.

[23] HARRINGTON, J., PALETHORPE, S. & WATSON, C.I.: Does the Queen speak the Queen's English? *Nature*, 408, pp. 927–928, 2000.

[24] HOLLIEN, H.: *The Acoustics of Crime: The New Science of Forensic Phonetics*. New York, London: Plenum Press, 1990.

[25] HOLLIEN, H.: An oilspill, alcohol and the captain: a possible misapplication of forensic science. *Forensic Science International,* 60(1–2), pp. 97–105, 1993.

[26] JACKSON, R.L.: Memo Helps Plug Gap in Watergate Puzzle: Scandal: The infamous erasure of an Oval Office tape recording by President Nixon's secretary was no accident, a newly released document suggests. *The Los Angeles Times,* Jan. 13, 1995.

[27] JOHNSON, K., PISONI, D. B. & BERNACKI, R.H.: Report to the NTSB: Analysis of speech produced by the captain of the Exxon Valdez. *Research on Speech Perception*, 15, pp. 1–9, 1989.

[28] JOHNSON, K., PISONI, D.B. & BERNACKI, R.H.: Do voice recordings reveal whether a person is intoxicated? A case study. *Phonetica,* 47(3–4), pp. 215–237, 1990.

[29] KERSTA, L.G.: Voiceprint Identification. *Nature*, 196, pp. 1253–1257, 1962.

[30] KERSTA, L.G.: Voiceprint Identification Infallibility. *The Journal of the Acoustical Society of America*, 34(12), p. 1978, 1962.

[31] KOENIG, W.: Final Report on Project C - 43. Continuation of Decoding Speech Codes. Part I – Speech Privacy Systems – Interception, Diagnosis, Decoding, Evaluation. New York: Bell Telephone Labs Inc., 1944.

[32] KOENIG, B.E.: Authentication of forensic audio recordings. *Journal of the Audio Engineering Society,* 38(1–2), pp. 3–33, 1990.

[33] KOENIG, B.E. & LACEY, D.S.: Forensic authentication of digital audio recordings. *Journal of the Audio Engineering Society*, 57(9), pp. 662–695, 2009.

[34] KOPP, G.A. & GREEN, H.C.: Basic Phonetic Principles of Visible Speech. *The Journal of the Acoustical Society of America*, 18(1), pp. 4–89, 1946.

[35] Krech, E.-M., Stock, E., Hirschfeld, U. & Anders, L.C.: *Deutsches Aussprachewörterbuch*. Berlin/New York: de Gruyter, 2009.

[36] Künzel, H., Braun, A. & Eysholdt, U.: *Einfluß von Alkohol auf Sprache und Stimme*. Heidelberg: Kriminalistik-Verlag, 1992.

[37] Maher, R. C.: Audio forensic examination. *IEEE Signal Processing Magazine*, 26(2), pp. 84–94, 2009.

[38] McGehee, F.: The reliability of the identification of the human voice. *Journal of General Psychology*, 17, pp. 249–271, 1937.

[39] McGehee, F.: An experimental study of voice recognition. *Journal of General Psychology*, 31, pp. 53–65, 1944.

[40] Pisoni, D.B., Yuchtman, M. & Hathaway, S.N.: Effects of alcohol on the acoustic-phonetic properties of speech. *Society of Automotive Engineers Technical Paper. (No. 860361)*, 1986.

[41] Pisoni, D.B., Johnson, K. & Bernacki, R.H.: Effects of alcohol on speech. In: *Proceedings of the Human Factors Society Annual Meeting*, 35(10), pp. 694–698. Los Angeles, CA: SAGE Publications, 1991.

[42] Potter, R.K.: *Field Uses of Voice Identification*. Memo of 31 May, 1944.

[43] Potter, R.K.: Visible Patterns of Sound. *Science*, 102(2654), pp. 463–470, 1945.

[44] Potter, R.K.: Introduction to Technical Discussions of Sound Portrayal. *The Journal of the Acoustical Society of America,* 18(1), pp. 1–3, 1946.

[45] Potter, R K., Kopp, G. A. & Green, H.: *Visible Speech*. New York: Dover Publications, 1947.

[46] Ray, M.: *The trial of the century* https://www.britannica.com/event/Lindbergh-baby-kidnapping/The-trial-of-the-century

[47] Ripley, A.: An expert hints at tape accident. *The New York Times*, 19 Feb, p. 1, 1974.

[48] Siebs, T. (Ed.): *Deutsche Bühnenaussprache*. Berlin/Köln/Leipzig, 1898.

[49] Tanford, J.A., Pisoni, D.B. & Johnson, K.: Novel scientific evidence of intoxication: Acoustic analysis of voice recordings from the Exxon Valdez. *The Journal of Criminal Law & Criminology*, 82(3), p. 579, 1991.

[50] Trouvain, J.: Notes on the development of speaking styles over decades – the case of live football commentaries. In: R. Hoffmann & J. Trouvain (Eds.): *HSCR 2015. Proceedings of the First International Workshop on the History of Speech Communication Research.* Dresden, September 4–5, 2015. Dresden: TUDpress, pp. 160–166, 2015.

[51] Trouvain, J. & Malisz, Z.: Inter-speech clicks in an Interspeech keynote. In: *Proceedings of Interspeech 2016.* International Speech Communication Association, pp. 1397–1401, 2016.

[52] Trouvain, J. & Truong, K.P.: Laughter. In: *The Routledge Handbook of Language and Humor*. Routledge, pp. 340–355, 2017.

[53] van den Berg, J.: Myoelastic-aerodynamic theory of voice production. *Journal of Speech and Hearing Research,* 1(3), pp. 227–244, 1958.

[54] Viëtor, W.: *Die Aussprache des Schriftdeutschen. Mit dem Wörterverzeichnis für die deutsche Rechtschreibung zum Gebrauch in den preussischen Schulen in phonetischer Umschrift sowie phonetischen Texten.* Zweite umgearbeitete Auflage der Schrift "Die Aussprache des Wörterverzeichnisses für die deutsche Rechtschreibung zum Gebrauch in den preussischen Schulen". Leipzig: Reisland, 1890.

[55] VON BÉKÉSY, G. & WEVER, E. G.: *Experiments in Hearing*. New York: McGraw-Hill, 1960.

[56] WELLS, J. C.: *Accents of English. Vol. II: The British Isles*. Cambridge: CUP, 1982.

[57] WOLF-HEIDEGGER, G. & CETTO, A.M.: *Die anatomische Sektion in bildlicher Darstellung.* Basel: Karger, 1967.

[58] YARMEY, A.D., YARMEY, M.J. & TODD, L.: Frances McGehee (1912–2004): The first earwitness researcher. *Perceptual and Motor Skills*, 106(2), pp. 387–394, 2008.

[59] ZENNER, H.-P., ARNOLD, W. & GITTER, A.H.: Outer Hair Cells as Fast and Slow Cochlear Amplifiers with a Bidirectional Transduction Cycle. *Acta Oto-Laryngologica*, 105(5-6), pp. 457–462, 1988.

[60] https://de.wikipedia.org/wiki/Entf%C3%BChrung_von_Ursula_Herrmann

[61] https://www.fbi.gov/history/famous-cases/lindbergh-kidnapping

HONOURING HISTORICAL FACTS: THE CASE OF INTONATIONAL DOWNTRENDS

Jan Volín, Pavel Šturm

Institute of Phonetics, Faculty of Arts, Charles University, Czech Republic
jan.volin@ff.cuni.cz

Abstract: Various researchers of the past are honoured in citations as the very first who described or discovered certain scientifically important phenomena. However, the illusion of "our times being superior to past" causes certain myopia: we often do not go back enough to identify and appreciate the true pioneers. Generally, we tend to underestimate what our ancestors already knew and did, and we overestimate the contribution of our own era. The present paper demonstrates this predicament with the case of intonational downtrends, i.e., gradual lowering of phonologically equivalent melodic targets within an utterance. The first report of the phenomenon is often attributed to K. L. Pike, sometimes even to later intonologists. We, on the other hand, document much older accounts of intonational downtrends, and offer some tentative explanations as to why the earlier research was ignored.

1 Introduction

1.1 Rationale

Science as a way of human learning about the world possesses several crucial attributes. One of them is cumulativeness. Researchers who expand human knowledge always build on what has been achieved before them.

Many lay people believe that in terms of scientific endeavour our ancestors were somehow underdeveloped or even backward. This cognitive illusion stems from several sources. First, it is the state of technology that leads to a childish contention like "our ancestors did not have our 'wonderful gadgets', therefore, they must have been intellectually deficient". This way of reasoning not only confuses innovations with science, but also ignores the fact that innovations largely benefit from science, and share the *cumulativeness* feature with it. Second, when teaching children to be grateful, there is a general tendency to portray the current state of affairs as advanced in comparison with backslidden or even degenerated past. Third, it is often pointed out how the outdated scientific views of the past were mistaken, but we do not realize that our current views will also be judged as outdated and mistaken one day. Moreover, what seems clearly wrong from today's perspective in the science of the past actually constitutes the essential precondition of our current knowledge, so even the errors should be considered with gratitude and respect. It follows that one does not have to be a trained philosopher to appreciate the past as the essential prerequisite of the present.

Current conventions of empiricist discourse require accurate and unambiguous referencing whenever non-trivial facts, propositions, assertions, or quantitative results are presented in research reporting (see, e.g., American Psychological Association [1]). This seemingly simple requirement causes a surprisingly large number of problems. Our present paper demonstrates one of them: attributing certain suggestions, findings or primacy to the wrong person. The reasons for this can be many – anything from good faith in a source, through plain ignorance up to ill intentions.

It has to be stressed, however, that our chief objective in the present study is not to criticize blunders of our colleagues, because we realize only too well that no one is really immune to slip-ups. Instead, we wish to expound some interesting facts about a feature of speech melody description that has been known for a longer time than various authors usually report.

1.2 Intonation and downtrends

Intonation itself was largely outside the focus of phonetic research until about 1960s. Although there were many inspired works from long before that (e.g., [10], [25], [2]), speech prosody was overshadowed by segmental issues. This can be easily documented by the lists of topics at the first three International Congresses of Phonetic Sciences (1932 in Amsterdam, 1935 in London, 1938 in Ghent), where papers related to melody of speech occupied an extremely small part of the space. Rossi [28: 2] speaks about an "exhaustive bibliography covering the years from 1900 to 1973" in which Di Cristo collected almost 4400 studies on speech prosody. When Rossi ordered them by their dates and split in the middle, he found out that the first half of them needed over 60 years to be written, while the other half was generated within just over 6 years. The turning point was 1967 (the year of the 6th ICPhS in Prague; see also [30] for similar comparisons between 1967 and 2011).

Various hypotheses could be stipulated concerning the unbalanced division of attention in the study of sound patterns in speech. First, it is true across all scientific disciplines that people initially deal with the obvious, and only later arrive at phenomena that are less accessible to conscious observation. Thanks to alphabetic spelling systems dominating the learned space in Europe and America, the segmental speech sounds, i.e., consonants and vowels, have always been the most salient constituents of the speech signal in terms of conscious accessibility. When, for instance, Kingdon [19: xxv] explains the difference between kinetic and static tones, he likens them to diphthongs and monophthongs as if he presumed that those would be better known to his readers.

Second, the linguistic theory of the early 20th century was deeply interconnected with semiotics – the general theory of signs. Since various prosodic units and features did not fit the dogma of a discrete and arbitrary sign, researchers were less keen on studying these. Even today, however, no agreement has been reached on the essential nature of the linguistic sign.

Third, the heavy burden of exclusive anthropocentrism in the early stages of linguistic endeavour prevented scholars from seeing language in its entirety. Rather than realizing that language is a complex system that enables communication of a wide range of meanings, conservative linguists limited their object of inquiry to lexicon and grammar. Various fundamental parts of the systems were labelled 'paralinguistic' and left to be studied by 'someone else'. Yet, there were voices already in the 1930s that the representational, affective and applicative components of the meaning form a unity that must be taken seriously [5]. (It is natural that a linguist must specialize in, let's say, morphonology, but to say that *only* this auxiliary discipline is the true linguistic undertaking would be deeply damaging to the whole discipline.)

It is becoming clearer and clearer nowadays that various features of intonation generate meanings of utterances the same way lexical items do, i.e., through interaction with contexts. Just like lexical items, intonational features possess potentials of meanings, and only through interacting with the concrete layers of the context the real meaning or sense is produced. Thus, intonation is not 'added' to words. It is *selected together with the words*, and it is uttered to interact with the context and to generate the intended meaning. No features of intonation should be a priori excluded from the language structure just because the current state of knowledge is unable to see their role in communication [22: 323].

As stated above, these ideas began entering the general approaches to speech communication extensively in the mid-1960s. At about the same time larger attention was becoming oriented towards a specific feature of F_0 contours: a gradual downtrend of F_0 values throughout utterances. This trend can be expressed as $\Delta F_0/\Delta t < 0$, and illustrated with linear regression lines throughout F_0 contours (e.g., Figure 1). More often, however, it is emphasized that the trend should be studied as the gradual lowering of the *equivalent melodic events* like peaks or valleys. Rather than linear regression fit, researchers may use toplines (connecting peaks), baselines (connecting valleys), or even simple curves constructed with prescribed procedures.

The first significant public presentation of the phenomenon to larger audiences seems to have happened at the 5th International Congress of Acoustics in 1965 in Belgium (as alleged by 't Hart et al. [33: 123]). A group of Dutch phoneticians affiliated with the Institute for Perception Research (IPO) in Eindhoven suggested a name for the phenomenon: they called it *declination*. They also demonstrated its perceptual relevance in Dutch utterances. Further research has added similar concepts under the labels such as downstep, downdrift, final lowering, etc. The umbrella term *downtrends* was suggested by Connell and Ladd [12: 2] to the best of our knowledge.

Current intonology accepts the presence of intonation downtrends as a feature of melodic forms in most languages of the world. John Ohala even states that perhaps all languages in the world might exhibit some form of downtrends [23: 31], and Hirst and Di Cristo [16: 18nn] suggest that downtrends need to be described in any comprehensive account of a language intonation structure.

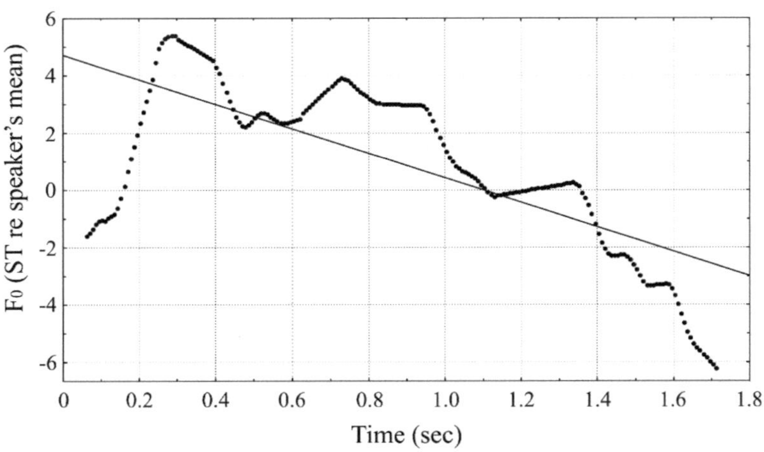

Figure 1. An illustration of an F_0 downtrend captured by a linear regression line. The utterance is part of a continuous narrative and says: "Zopakovala jsem to ještě jednou." ("I repeated it once again.")

When studying the past accounts of the downtrends, we noticed that while the terminology or methods of measuring and expressing them differed immensely, quite a few papers were in agreement about the fact that the first person to report downtrends was Kenneth L. Pike in 1945. Our key question in the present paper then asks whether the downtrends were really noticed only as late as in the 1940's.

2 Credits due?

2.1 Kenneth L. Pike

The American linguist and polyglot Kenneth L. Pike influenced quite profoundly several areas of linguistics and phonetics. Many authors hint, but some even allege that the first published account of downtrends in intonation was given by Pike in his book *The Intonation of American English* (1945). We will now list the suggestions in that respect and we will discuss their explicitness. The quotes are ordered by the date of their publication.

[9: 254] = Cohen et al., 1982

"*The phenomenon to which we want to refer, declination, was first reported by Pike*"

This reference is straightforward and unambiguous. In their paper for an international peer-reviewed journal, the authors do not seem to express any doubt about the fact (and they reassert it on p. 256). Since the paper is designed as an overview of the research area and a discussion of previous interpretations of the phenomenon, the claim sounds quite authoritative and may have been influential.

[33: 121] = 't Hart et al., 1990

"*In languages like Dutch and English, the most important global attribute is the observed tendency of F_0 to decrease slowly from beginning to end of an utterance. This phenomenon has already been reported by Pike.*"

This quote actually represents the same authors as the previous one, only in different order. It can be noted that eight years after the previous claim the wording is less definite here: Pike's primacy is only suggested. One can only speculate whether this is due to the general shift in academic discourse towards less absolute stipulations or whether the authors' confidence was weakened by some revelations.

[29: 212] = Schuetze-Coburn et al., 1991

"*The concept of declination, or gradual falling off of pitch during an utterance, was noted by Pike (1945) and Bolinger (1964) in perceptual data.*"

The authors do not explicitly claim that Pike was *the first*, yet they do not refer to anyone before. As the authors are Americans, we could perhaps understand why they name only two American linguists. This kind of myopia is common to all of us.

[13: 2161] = Dodane et al., 1999

"*Pike was the first one who noticed the decrease of F_0 through time (drift).*"

The authors do not seem to hesitate at all – the statement opens the section "Theoretical Elements" in their paper. They do not state where their certainty is coming from, they seem to know. It has to be emphasized here that we do not intend to criticize the authors. Instead, we want to demonstrate how easily all of us are led.

[24: 2] = Paeschke and Sendlmeier, 2000

"*Declination was first reported by Pike: The general tendency of the voice is to begin on a moderate pitch and lower the medium pitch line during the sentence.*"

The same case as the previous one: there seems to be no doubt on the part of the authors about the claim. The verbatim quote suggests that the impressive Pike's monograph was at the authors' disposal, which may have influenced their decision to nominate Pike as the discoverer.

[15: 537] = Hird & Kirsner, 2002

"*Declination is categorized as one of the global features of the prosodic contour ... and refers to the perceived lowering of fundamental frequency across a spoken utterance (Pike, 1945).*"

The form of the reference seems to lead to a single interpretation, but Pike speaks neither of fundamental frequency nor utterance. The authors most probably adapted the idea from a secondary source. It is also possible that they treat their sources with greater freedom than is common in current scientific discourse or that they do not differentiate between pitch and fundamental frequency or sentence and utterance. As to our chief concern at this point, the wording of the reference to Pike does not constitute a strong claim about his 'prerogative'.

[20: 75] = Ladd, 2008

> "*It has been clear at least since Pike (1945) that* F_0 *tends to decline over the course of phrases and utterances, both in tone languages and in languages like English or Dutch.*"

This is an opening sentence of a section in a book printed by a prestigious British publisher. The author quite probably understands the weight vested on the claims in these initial positions, which could perhaps explain why he is careful with the stipulation and leaves the back door open with the expression *at least since*. Alternatively, he may have already encountered allusions concerning possible earlier accounts. Be that as it may, the book is a very frequently cited source, so its readership is considerably large. Interestingly, the author speaks briefly about downtrends already on p. 16 in introductory remarks, but does not go before the year 1967 in references there [20: 16].

2.2 Indistinct primacy

A large body of research in the area of downtrends has been produced over the last decades. Many authors actually avoid the considerations concerning the first person to report the phenomenon. They nevertheless refer to the past research as required by contemporary reporting conventions, but they stay less explicit. Although such citations are not directly binding, they might be potentially misleading. Our following examples are ordered by the year which is insinuated as initial in downtrend research.

Patterson [26: 25] in his dissertation introduces the concept of declination by reference to Liberman & Pierrehumbert [21], although two pages later he goes back to 1967 to give credit to the IPO researchers, even if they actually introduced the concept of declination two years before that (cf. Cohen & 't Hart [8]).

Liberman and Pierrehumbert themselves dedicate major space to downtrends in their extensive and often cited study, yet when it comes to the history of declination research, they do not go further back than to 1969 [21: 163, 224, 232 – note 1].

Ohala, who was known for his vast knowledge of literature even before the era of the internet, produced an ample reference section for his comprehensive overview of tone production, where he cites even an author from the 16th century. However, when he discusses downtrends, he does not go further back than to 1970 [23: 31, 32].

Clark and Yallop produced a best-selling university textbook of the foundations of phonetics and phonology. It is used as a reliable compendium by many scholars in linguistics and related disciplines. However, in their treatment of downtrends they do not go before the year 1975 [7: 284-286].

Equally short 'memory' can be also found in the study by Terken [31]. Although declination is a key term in that study, his initial references do not reach beyond the above-mentioned influential study of Liberman & Pierrehumbert from 1984. Judging from his concise style of writing, it is perhaps possible that he liked their simple wording in the definition [31: 1769]. However, in the General Discussion the fact that he belonged to the IPO team is reflected in his reference to the work of his older colleagues in 1975 [31: 1775], referring to 't Hart and Collier [32].

A very special case is that of Roger Kingdon [19], whom we could not resist to add to this section. In his widely used monograph, Kingdon introduces the phenomenon quite clearly:

> "*When a series of level tones occurs in the head and body of a tune, each succeeding level tone is pitched a little below its predecessor; a long sentence consists, therefore, of a slowly descending series of stressed syllables interspersed with unstressed syllables which remain on or very close to the pitch of the stressed syllables...*" [19: 25]

On the same page the author emphasizes that slowly descending stressed syllables are "a basic feature of English intonation". However, he does not refer to any other authors who would corroborate this view. By doing so, Kingdon presents the downtrend as something common and undisputable.

2.3 Earlier accounts

Certain doubts about Pike's primacy were expressed cautiously by Volín [34: 2], who knew the drawings in the monograph by Armstrong and Ward from 1926 (see below). Only later, he discovered that his doubts were substantiated by the research of Collins and Mees [11], who actually suggested to go even before 1926. This directed us to the oldest source cited in the present paper: *Intonation Curves* (1909) by Daniel Jones [17]. In this extraordinary work Jones mapped an unprecedented amount of speech material in three languages: English, French and German. All the samples represent acted speech (drama, poetry, acted conversation), and Jones depicts the tunes of utterances with curved line drawings in a musical stave to "combine as far as possible scientific accuracy with practical utility" [17: v]. The curved lines were made through perceptual analysis over repeated playback of short stretches from a gramophone record. The drawings start appearing on p. 3, where there are 10 utterances, of which 8 exhibit a downtrend. An example is displayed in Figure 2.

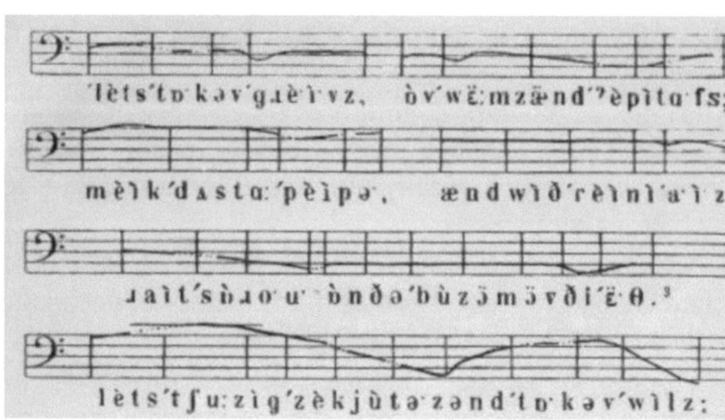

Figure 2. An example from Jones [16: 3] with moderate to steep downtrends.

Collins and Mees in their thorough biography of Daniel Jones refer to hand-written lecture notes in which the English phonetician illustrates downstep of lexical tones in Sechuana (i.e., Tswana) [11: 161]. They document that a couple of years later, Jones presented the downstep in Sechuana in print, specifically in *Transactions of the Philological Society* [11: 187].

One of the handbooks in long-term high demand was Jones' *Outline of English Phonetics*, which was first published in 1918. Its 9th edition was published in 1960 and reprinted several times, also in 1969, two years after the author's death. Jones kept continuously rewriting parts of the text, improving the transcriptions, and adding materials into the book. The oldest edition that we were able to access was the second one [18]. The chapter on intonation there stretches over 34 pages and Jones uses musical notation together with curved lines to capture the pitch tracks. Many of the examples in the *Outline* are actually too short to become the domain of a downtrend, but the longer ones show it unambiguously. More importantly, Rule 10 in the chapter says:

> "*When in a sentence having a falling intonation there are a number of stressed syllables, the first important stressed syllable generally has the highest tone and the other important syllables form a descending series of notes.*" [18: 161]

Harold Palmer is another author who represents what is referred to as the British school or British tradition (e.g., [14: 278] or [14: 301], respectively), and he is also clearly aware of the downtrends. The evidence can be found in his highly praised handbook from 1922 both in some of the diagrams of the tunes there and in the wordings such as: "the first syllable of a superior head is on the high pitch, each successive syllable being pitched slightly lower than the one before" [25: 45]. It is interesting to see that Palmer depicts the downtrend as independent from the final nucleus. Figure 3 presents three different nuclei, each of which is preceded by descending intonation.

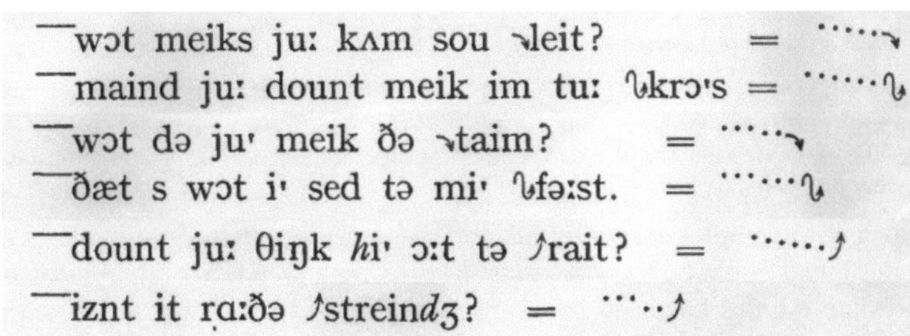

Figure 3. Descending tunes as captured by Palmer [25: 45].

Four years after Palmer and eight years after the first edition of the *Outline*, two members of the Department of Phonetics at University College London, Armstrong and Ward, published a more extensive and detailed treatise on English intonation [2]. Interestingly, the book is dedicated in capital letters "TO PROFESSOR DANIEL JONES" (Figure 4), but at the same time, Jones gives credits to both authors in his later works and adopts many of their improvements in his own intonological studies. This can be interpreted as an indication of bidirectional influence and gratitude. The book was accompanied by three double-sided gramophone records which comprised Armstrong and Ward's readings of English passages, and was in print for about 50 years. It inspired a large number of intonologists over the world, and even its critics mostly just repeated what the authors themselves already admitted and defended as their intention in the context of their work.

More to the chief concern of our study, numerous tunes in the book clearly contain undisputable downtrends and the very first example of English intonation on p. 4 is no exception. Right under this first example, there is a sentence: "The stressed syllables form a descending scale." This assertion is apparently meant as a general rule.

One of the early published reviews of the book was written by Anna Bohnhof [3]. Although the reviewer does not explicitly mention intonation downtrends in any sort of a defining statement, the illustrations of tunes reprinted from Armstrong and Ward (1926) clearly contain them. Their salience is indisputable (Figure 5).

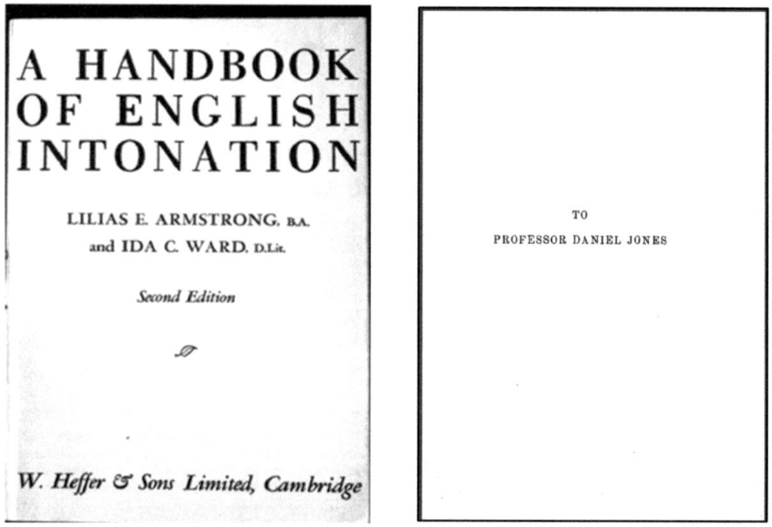

Figure 4. The opening page and the dedication from the second edition of Armstrong & Ward's principal work on intonation [2].

For some reason, the examples are reprinted in the review without the guiding lines demarcating the pitch range so they might be less convenient for reading than the original ones. Nonetheless, 8 out of 16 example tunes have the structural potential to host downtrends, i.e., at least two stressed syllables before the nuclear pitch accent. All of these materialize the potential and display the downtrend. Moreover, one of the tunes demonstrates the so-called declination reset, which is commented upon in the text [3: 270].

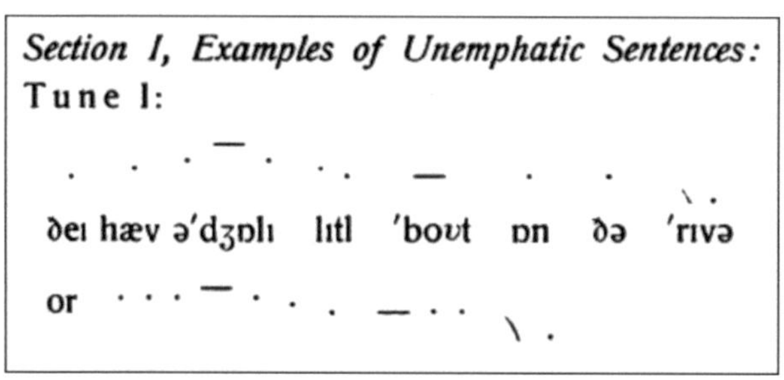

Figure 5. An illustration from [3: 270] with a clear display of the downtrend.

Still ten years before Pike, there is another interesting evidence of downtrend awareness clearly documented in the contribution of S. Boyanus to the 2nd International Congress of Phonetic Sciences in 1935 [4]. The Congress took place in London, where Boyanus settled permanently a year before that, obviously after marrying Lilias Armstrong. His presentation concerned principal sentence melodies of Russian – his mother tongue. He used 28 stave illustrations of melodies, of which 8 were long enough and structurally suited for downtrends to unfold in them. Indeed, all eight of them do display downtrends (as an example see Types 1 and 3 in Figure 6).

However, the peculiar point here is that Boyanus never comments upon the downtrend in the conference paper. In a footnote, though, he acknowledges the influence of his wife and her colleague Ida Ward on his work. Therefore, it is quite possible that the clear presence of downtrends in the illustrations was their contribution, while Boyanus himself considered the phenomenon perhaps less important to talk about.

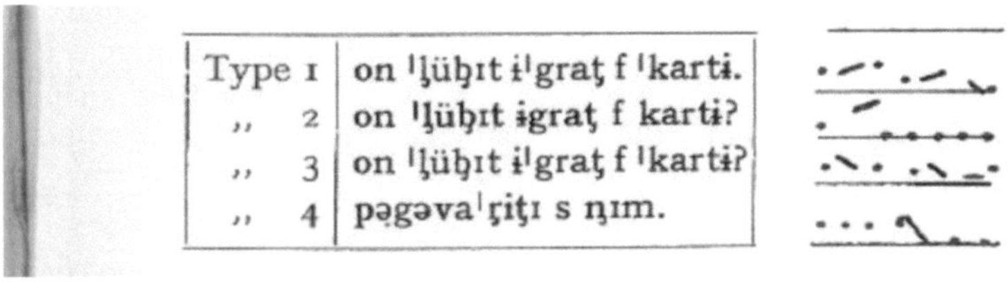

Type 1	on ˈl̡üb̡ɪt ɨˈgrat̡ f ˈkartɨ.
,, 2	on ˈl̡üb̡ɪt ɨgrat̡ f kartɨ?
,, 3	on ˈl̡üb̡ɪt ɨˈgrat̡ f ˈkartɨ?
,, 4	pəgəvaˈr̡it̡ɪ s ŋɪm.

Figure 6. Four examples from Boyanus [4: 111], of which 2 are structurally apt to display downtrends.

3 Discussion

We have demonstrated in our survey that downtrends were noted and reported long before it is commonly acknowledged. Why is the work of D. Jones and his colleagues L. Armstrong and I. Ward not referenced? The hypothesis that it might be unknown can only be applied to Jones' notes from about 1916 reported in [11]. Both Jones' *Outline* and Armstrong and Ward's *Handbook* were very well known before the often-cited Pike's statement from 1945. Could the problem then stem from the fact that Pike's account is longer, more thorough or more sophisticated? Not at all, since it is not. Pike chooses the label *drift* for the downtrend and explains it with very simple words in two sentences [27: 77]. He mentions the term briefly in a few occasional remarks later in his book without any additional explanations.

When looking for alternative explanations, we might also speculate about the standing of the American structuralist tradition and the didactically oriented efforts of the London school. Jones and his colleagues never denied their ambitions to help language teachers. As Fox characterizes them: "Their concern is less with theoretically adequate formalization than with the identification of pedagogically useful categories" [14: 127]. Since there used to be, and sometimes still is, certain prejudice against pedagogically oriented research, perhaps writers were shy to refer to sources that might be considered less serious. That would be clearly wrong, and not only because history showed that Jones was in the end extremely influential even without being 'complicated'.

Nevertheless, the illusions mentioned in the introduction of our paper together with certain carelessness might be the most plausible explanation. The error of *present-centeredness* of historical accounts was exposed already in 1931 in a brilliant essay by Herbert Butterfield [6], and both historians and philosophers keep elaborating on this serious cognitive bias, knowing that it is omnipresent and difficult to eradicate (e.g., [35]).

In our paper we would like to warn against compulsory historical precision, which would be unattainable, but also against careless referencing to whatever is at hand. Human behaviour shows that despite all the clues we receive from the surrounding world, inclination to extremes is often the prevailing modus operandi. Despite all the resources that are available nowadays thanks to the internet, we cannot obligate authors to know the history of their discipline in every detail. On the other hand, there should be very low tolerance towards haphazard referencing, let alone crediting friends and colleagues for trivial claims of the type "speech is important in our everyday lives".

We have contributed to the question of historical accounts of intonational downtrends. Apart from the possibility of even earlier observations than reported here, our survey leaves some further questions unanswered. For instance, a thorough research would be needed to find out who actually first appreciated important *functions* of downtrends in various languages or speaking styles. Similarly, it might be interesting to know who first perceptually tested or at least pointed out the contribution of downtrends to naturalness of the speech signal.

Finally, both experts and laymen may ask whether it really matters who the first discoverer was. The answer would depend on the framework of values in which we choose to operate. If we measure progress by the number of new products that can be bought by consumers, then it does not matter. If, however, we measure progress by our respect for the world, then it does matter. We can only truly admire this world and have faith in it if we understand it without bias.

Acknowledgments

This research was supported by the Charles University project *Progres Q10: Language in the shiftings of time, space, and culture*.

References

[1] AMERICAN PSYCHOLOGICAL ASSOCIATION. *Publication Manual of the American Psychological Association [7th Edition]*. APA, 2020.

[2] ARMSTRONG, L. E. & WARD, I. C.: *Handbook of English Intonation*. Cambridge: Heffer, 1926. [2nd edition printed in 1931 was used here]

[3] BOHNHOF, A.: Review of Handbook of English Intonation by Lilias E. Armstrong and Ida C. Ward. *Neuphilologische Mitteilungen,* 29(4), pp. 269–272, 1928.

[4] BOYANUS, S. K.: The main types of Russian intonation. In: *Proceedings of 2nd ICPhS*. IPA: London, pp. 110–113, 1935.

[5] BÜHLER, K.: *Sprachtheorie. Die Darstellungsfunktion der Sprache*. Jena, Fischer, 1934. — *Theory of Language. The representational function of language* [translated by D. F. Goodwin in collaboration with Achim Eschbach]. Amsterdam: John Benjamins, 2011.

[6] BUTTERFIELD, H.: *The Whig Interpretation of History.* New York: W.W. Norton & Company, 1965. [first published in 1931]

[7] CLARK, J. E. & YALLOP, C.: *An Introduction to Phonetics and Phonology*. Blackwell, 1990.

[8] COHEN, A. & 't HART, J.: Perceptual analysis of intonation patterns. In: D.E. COMMINS (Ed.): *Proceedings of the 5th International Congress on Acoustics*. Liège, Sept. 7–14, 1965. Liège: Imprimerie G. Thoné, paper A16, 1965.

[9] COHEN, A., COLLIER, R. & 'T HART, J.: Declination: Construct or intrinsic feature of speech pitch? *Phonetica*, 39(4/5), pp. 254–273, 1982.

[10] COLEMAN, H. O.: Intonation and emphasis. *Miscellanea Phonetica I*. International Phonetics Association, pp. 6–26, 1914.

[11] COLLINS, B. & MEES, I. M.: *The Real Professor Higgins: The Life and Career of Daniel Jones*. Berlin: De Gruyter Mouton, 1998.

[12] CONNELL, B. & LADD, D. R.: Aspects of pitch realisation in Yoruba. *Phonology*, *7*(1), pp. 1–29, 1990.

[13] DODANE, C., KONOPCZYNSKI, G., SANTI, S. & ESPESSER, R.: F0 declination line: More evidence from infants and children's speech (9-48 months). In: *Proceedings of the 14th ICPhS*. San Francisco: UCLA, pp. 2161–2164, 1999.

[14] FOX, A.: *Prosodic Features and Prosodic Structure*. Oxford: Oxford University Press, 2000.

[15] HIRD, K. & KIRSNER, K.: The relationship between prosody and breathing in spontaneous discourse. *Brain and Language, 80*(3), pp. 536–555, 2002.

[16] HIRST, D. & DI CRISTO, A. (1998): *Intonation Systems*. Cambridge: Cambridge University Press, 1998.

[17] JONES, D.: *Intonation Curves*. Leipzig: Teubner, 1909.

[18] JONES, D.: *Outline of English Phonetics* [2nd edition]. New York: G. E. Stechert & Co., 1922.

[19] KINGDON, R.: *The Groundwork of English Intonation*. London: Longman, 1958.

[20] LADD, D. R.: *Intonational Phonology* [2nd edition]. Cambridge: Cambridge University Press, 2008.

[21] LIBERMAN, M. & PIERREHUMBERT, J.: Intonational invariance under changes in pitch range and length. In: M. ARONOFF & R. T. OEHRLE (Eds.): *Language Sound Structure*. Cambridge, MA: MIT, 1984.

[22] LOCAL, J.: Variable domains and variable relevance: Interpreting phonetic exponents. *Journal of Phonetics*, 31, pp. 321–339, 2003.

[23] OHALA, J. J.: Production of tone. In: V. A. FROMKIN (Ed.): *Tone: A Linguistic Survey*. New York: Academic Press, 1978.

[24] PAESCHKE, A. & SENDLMEIER, W. F.: Prosodic characteristics of emotional speech: Measurements of fundamental frequency movements. In: *Proceedings of ISCA Workshop on Speech and Emotion*. Belfast: ISCA, (without page numbers), 2000.

[25] PALMER, H. E.: *English Intonation with Systematic Exercises*. Cambridge: Heffer, 1922.

[26] PATTERSON, D.: *A Linguistic Approach to Pitch Range Modelling*. Unpublished doctoral dissertation, University of Edinburgh, 2000.

[27] PIKE, K. L.: *The Intonation of American English*. Ann Arbor: University of Michigan Press, 1945.

[28] ROSSI, M. (1997): Intonation: past, present, future. In: *Intonation: Theory, Models, Applications*. Athens: ESCA, pp. 1–10, 1997.

[29] SCHUETZE-COBURN, S., SHAPLEY, M. & WEBER, E. G.: Units of intonation in discourse: A comparison of acoustic and auditory analyses. *Language and Speech*, 34(3), pp. 207–234, 1991.

[30] ŠTURM, P.: International phonetic congresses: The shift in research practices and areas of interest over 44 years. In: *Proceedings of the 18th International Congress of Phonetic Sciences*. Glasgow: University of Glasgow, paper 182, 2015.

[31] TERKEN, J. M. B.: Fundamental frequency and perceived prominence of accented syllables. *Journal of the Acoustical Society of America*, 89, pp. 1768–1776, 1991.

[32] 't HART, J. & COLLIER, R.: Integrating different levels of intonation analysis. *Journal of Phonetics*, 3, pp. 235–255, 1975.

[33] 't HART, J., COLLIER, R. & COHEN, A.: *A Perceptual Study of Intonation: An Experimental-Phonetic Approach to Speech Melody*. Cambridge: Cambridge University Press, 1990.

[34] VOLÍN, J.: *Downtrends in Standard British English Intonation*. Frankfurt am Main: Hector, 2008.

[35] WILSON, A. & ASHPLANT, T. G.: Whig history and present-centred history. *The Historical Journal*, 31(1), pp. 1–16, 1988.

Aerodynamic voicing constraint in stops: from Wolfgang von Kempelen to John Ohala

Jürgen Trouvain

Saarland University, Language Science and Technology, Saarbrücken, Germany
trouvain@lst.uni-saarland.de

Abstract: Voicing in stop consonants is restricted by aerodynamical constraints, as illustrated by John Ohala, among others, more than three decades ago. It seems to be unnoticed in modern phonetic research that Wolfgang von Kempelen already described these challenges two centuries earlier. With additional bellows in his speaking machine he offered a mechanical solution for the vocal tract enlargement in stop production to achieve a longer voiced phase. A look back can help to develop a better understanding of the main mechanism for contrasting stops in many languages.

1 Introduction

The aim of this paper is to enlighten the non-trivial mechanism of voicing in stop consonants. This mechanism is strongly constrained by aerodynamics, which has been described in modern phonetics, mainly by Ohala [20, 21] in the last quarter of the 20th century, but also in historical sources, in particular by von Kempelen [14] at the end of the 18th century.

There is no doubt that there is a huge body of research on stops in the phonetic sciences in the last century or so. However, it does not seem to be the case that the voicing of stops is fully understood. For instance, investigations on the well-known phonological phenomenon of final devoicing in German ("Auslautverhärtung") reveal particular patterns of partial devoicing, e.g. [15]. In addition, the degree of voicing and devoicing plays a central role for the acquisition of the voiced-voiceless contrasts in stops and fricatives in second language learning [6].

A look at aerodynamical constraints on stop voicing seems also promising when attempting to model voiced stops. In addition to transcription-based auditory phonetics and acoustics-based instrumental phonetics, the modelling of concrete sound events provides extra information on the details of speech production. This phonetic modelling can for instance be performed through software-based articulatory synthesis [3] or through hardware-based experiments, e.g. [4]. The historical descriptions by von Kempelen have the advantage that replicas of his speaking machine offer mechanical models for investigating the aerodynamic voicing constraints on stops.

2 Voiced-voiceless contrasts in stops

Most languages of the world have stop consonants in their sound inventories. Most sets of stops in those languages show a two-way distinction for different places of articulation, though there are three-way distinctions as in Korean or Armenian or even four-way distinctions as in Hindi. Among the two-way distinctions the predominant set is to have the contrasts /p-b/, /t-d/ and /k-g/. The term voiced vs. voiceless is mainly used to characterise this phonological contrast. It must be noted that the distinction in voicing is often used as an umbrella term for linguistic categorisation but that on the phonetic side voicing is not, or only partially, used to maintain the contrast in some languages.

There are alternative ways to distinguish between the phonologically voiced and unvoiced stops. This includes differences in voice onset time (*VOT*) and *aspiration*, e.g. in English and many other languages, see e.g. Adjarian [1, 8] for very early investigations and some overviews such as Lisker & Abramson [17] and Cho & Ladefoged [10]. For German this contrast is often called lenis-fortis instead of voiced-voiceless [7, 12]. In addition, there is a tendency among many languages, not only German, to *final devoicing*, as Keating et al. [13] show.

There are also cases where stops contrast only in *quantity* and where aspiration and voicing play no role. For instance, in many Swiss German varieties there are differences in the duration of unvoiced closure phases between phonologically voiced and unvoiced plosives [16]. Also Ohala [21] observed a general tendency across languages for longer duration in voiceless obstruents compared to voiced ones.

Another example of not using voicing of the stops is the effect of *pre-consonantal vowel lengthening* (also called pre-consonantal voicing effect). It refers to vowels preceding phonologically voiced obstruents being consistently longer than their phonologically voiceless counterparts, such as the differences in the English minimal pairs 'bad-bat' and 'phase-face' [11, 24].

Voiced obstruents seem to be less frequent than their voiceless counterparts in phoneme inventories of the world's languages [18, 27]. However, voiced types of obstruents are nonetheless relatively common with more than 60% in the data of the World Atlas of Language Structures [18, 19]. There is a remarkable number of languages that lack a /g/ in the consonantal inventory. Maddieson [18] concludes that "[b]ecause of the aerodynamic facts, /g/ can be seen to be a less favored plosive than /b/ or /d/."

3 Aerodynamic constraint on voicing in stops in human speech

3.1 Air pressure and phonation

Typically voicing ceases after 15 ms [20]. The closure phases of fully voiced plosives are usually considerably longer than that and it is assumed that the vocal tract is enlarged in order to maintain a trans-glottal flow and thus delay the cessation of voicing. The data in Ohala and Riordan [20] showed a duration of about 82 ms as the limit for voicing in labials and, as expected, much shorter (52 ms) in velars, if purely *passive* expansion of the vocal tract walls is considered. The authors estimate that each cm^3 gives an estimated additional 10 ms of voicing.

One of the prerequisites for phonation is a sufficiently large difference of the transglottal air pressure to maintain airflow. After the oral closure for a stop, supra-glottal air pressure increases and leads to a reduction of the transglottal air pressure drop. When the difference is not large enough anymore, devoicing applies.

Catford [9] estimates the threshold value of the pressure differential at about 2-3 cm H_2O with adducted vocal folds in a modal phonation setting. He further estimates the volumes contained between the glottis and the oral closure for different places of articulation as 30-50 cm^3 for velar stops, 70-100 cm^3 for alveolar stops, and 120-160 cm^3 for bilabial stops. For a [g] there is less volume than for [d] and [b] and it can be expected that the devoicing happens earlier, as the place of closure is closer to the glottis. This is the explanation behind the frequent absence of /g/ mentioned above.

3.2 Vocal tract enlargement

This *passive* enlargement of the vocal tract happens through the compliance of tissue [20, 21, 22, 23, 26]. An *active* expansion of the vocal tract can be achieved for example by lowering the larynx or by lowering the tongue body or by elevating the already closed soft palate a bit more, and expanding the pharyngeal walls [20, 21, 22, 23, 26].

4 Aerodynamic constraints in stops in Kempelen's speaking machine

4.1 Voicing in stops in human speech

The aerodynamic constraints in voiced stops were already observed and described by Wolfgang von Kempelen [14] two centuries earlier. In his famous book, chapter 4 is devoted to the speech production mechanisms of "the sounds or letters of the European languages". For the description of [b] he mentioned that the voiced phase has a limited duration [14, p. 242]:

"Man kann auch bey geschlossenem Munde, und Nase eine Stimme hören lassen, jedoch nur eine kurze Zeit, und nicht sehr laut."

"When the mouth and nose are closed, one can hear the voice, but only for a short time and not very loudly."

What follows is the explanation of the difference of the transglottal air pressure and the drop of this pressure difference needed for phonation after a short while [14, p. 243]:

"Wenn nun die Stimme ansprechen soll, so wird die in der Lunge enthaltene Luft zusammengedrückt, das Stimmhäutchen öffnet sich ein klein wenig, und gestattet ihr einen nur ganz engen Durchgang. In der in dem Munde vorräthigen, viel weniger zusammengedrückten Luft findet sie noch so viel Raum, daß sie in dieselbe mit einem Laut hineinströmen kann, indem die diese immer mehr und mehr zusammendrückt. So bald die Luft in dem Munde eben so sehr zusammengedrückt ist, als die in der Lunge, so ist zwischen beyden das Gleichgewicht hergestellt, der Strohm der Luft hört auf, und mit ihm die Stimme. Und dieses ist die Ursache, warum die Stimme nur eine kleine Weile, etwan 1 Sekunde lang anhalten kann."

"If now the voice wishes to speak, the air contained in the lungs is compressed, the vocal cords open a small amount, and allow it only a narrow passage. In the much less compressed air that is found in the mouth, it finds so much space that it can flow in, producing a sound, as it compresses the latter more and more. As soon as the air in the mouth is as compressed as that in the lungs, an equilibrium is established between the two, the stream of air ceases, and with it the voice. And this is the reason why the voice can only last for a short while, about one second."

He indicates an entire second for this aerodynamic manoeuvre of voicing in the (bilabial) stop, but such a time interval should be regarded as a rather short duration of auditory observation rather than a precise measurement that would be possible much later. Interestingly, he also offers the vocal tract enlargement in its passive and in its active form as an explanation of the underlying speech production mechanism [14, p. 244]:

"Nebst der Zusammendruckbarkeit der Luft bedienet sich die Natur noch eines anderen Kunstgriffes um der aus der Lunge kommenden Luft Raum zu verschaffen. Die Wände des ober dem Stimmhäutchen befindlichen Behältnißes, nämlich die fleischigen und daher nachgebenden Theile des inneren Halses erweitern sich, oder werden vielmehr von der anschwellenden Luft aufgebläht. [...] Wenn diese Theile einmal auf das höchste aufgetrieben sind, so muß die Stimme aufhören. Bläßt man über dieses noch die beyden Backen auf, so kann man die Stimme noch eine Weile länger aushalten."

"Besides the compressibility of the air, nature provides another artifice to make space for the air coming from the lungs. The walls of the chamber above the vocal cords, namely the fleshy and thus compliant portion of the inner throat expand, or rather are inflated by the air that is forced in. [...] When these parts are expands as far as they will go, the voice must cease. If one in addition expands the two cheeks, one can maintain the voice a bit longer."

4.2 Vocal tract enlargement in the speaking machine

Von Kempelen was also interested in modelling the observed effect of vocal tract enlargement. In his speaking machine [14] the vocal tract enlargement can be obtained by means of the 'plosive bellows'. These bellows are located directly beneath the 'nasal cavity', and the two cavities are linked with a small tube (Figures 1 and 2).

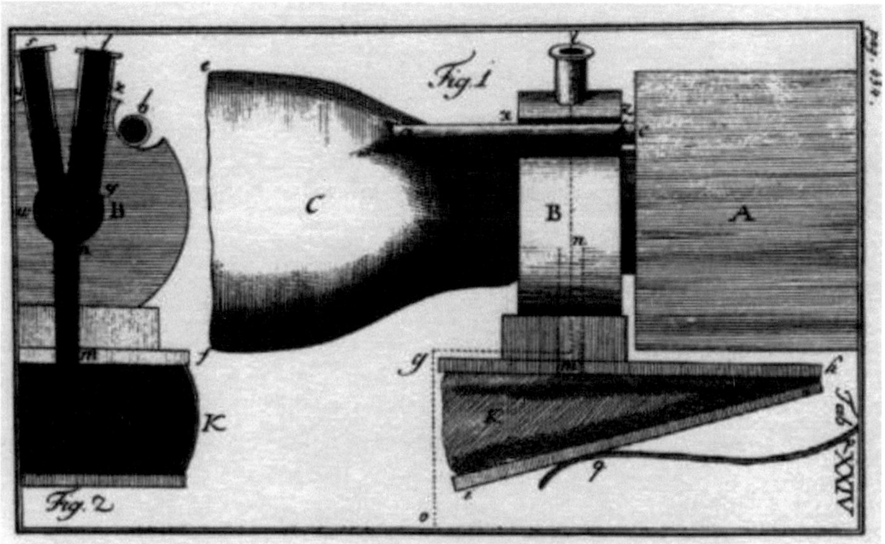

Figure 1. Plosive bellows "K" seen in cross-section in the front view (left-hand side) and in the side view of the inner life of the speaking machine (taken from the original engraving in Kempelen's book from 1791 [14]). The 'nasal cavity' ("B") is linked with a small tube ("n") to the plosive bellows located beneath.

These second, much smaller bellows also have the effect of lengthening the voiced phase of the plosives. There is evidence of this effect for various replicas. However, replicas without these bellows are better at generating voiceless plosives with an aspiration phase before voicing begins. An aspiration effect is not possible in replicas with these bellows, where a 'Papa' sounds more like a 'Baba' (at least to speakers of West Germanic languages). The possibility of switching the second bellows on and off would clearly offer a good solution to control the 'aerodynamic voicing constraint' [21].

It must be noted, however, that it was von Kempelen's belief that he was invoking exactly the opposite effect with the installation of the additional bellows [14, p. 437f.]:

"Um die Explosion bey den stummen Mitlautern zu verstärken, habe ich noch einen anderen, nicht minder wichtigen Zusatz gemacht. Ich habe nämlich an das Stück *B* unten einen kleinen Blasebalg *g h i* angebracht [...] Wenn nun Mund und Nase geschlossen ist, und der große Blasebalg gedrückt wird, so bläht die gepreßte Luft diesen kleinen Blasebalg mit auf. Wird sodann die Hand, die den Mund verschlossen hält, jäh von da weggerückt, so stoßt der kleine Blasebalg, der durch die an der Windlade befestigte Drahtfeder *p q* zusammen gedrückt wird, die Luft rasch von sich, wodurch die Explosion zum Munde hinaus nothwendig stärker wird. Und so haben wir nun den Buchstaben P in seiner Vollkommenheit."

"In order to strengthen the explosion with the voiceless consonants, I made yet another no less important addition. I installed underneath piece *B* a small bellows *g h i* [...] If the mouth and nose are closed, and the big bellows is pressed, the compressed air inflates this small bellows. If then the hand, which holds the mouth closed, is taken away, the small bellows, which is pressed together by the compression spring *p q* that is fastened to the wind chest, presses the air quickly out, by which means the explosion at the mouth is necessarily stronger. And thus we have the letter P in its full form."

Although the reasoning for generating an optimal [p] – instead of [b] – with the additional mechanical detail was not correct, it is remarkable that von Kempelen's observations and explanations on the aerodynamic voicing constraint on stops and the vocal tract enlargement were correct – a long time before modern and instrumental phonetics. In addition, his engineering approach of integrating small bellows for the stop production is admittedly sensational.

Figure 2. Side view of the inner life of the speaking machine of the Saarbrücken replica (measures in centimetres). Photography corresponds to the right-hand side illustration of Figure 1. Plosive bellows at the bottom are not inflated.

In order to obtain a distinction between [b] and [p] with the 'speaking machine', the player has to control the air pressure (by her/his elbow). The air pressure is increased for [p] compared to [b] immediately before and during the closure of both apertures. However, the initiation of 'subglottal' pressure must not be too strong, otherwise the onset of the 'vocal fold' vibration fails.

5 Considerations for future studies

The aerodynamic voicing constraint in stop consonants has been described by John Ohala, among others, more than three decades ago. It seems to be unnoticed in modern phonetic research that Wolfgang von Kempelen already described these challenges two centuries earlier. But still today it is unclear how exactly the enlargement of the vocal tract is executed for phonetically voiced stops.

A better understanding of the mechanism of voicing in stops can also help to explain further phenomena linked to this mechanism. One example refers to differences in how languages and language varieties use voicing in stops, maybe along with variation in VOT, closure duration and pre-consonantal vowel lengthening. Another example is variation in microprosodic details in, before, and after stops, that helps to improve articulatory synthesis [5]. A further example would apply to human non-verbal vocalisations such as crying but also screaming, with, so far, hardly any knowledge about vocal tract configurations, adjustments of vocal fold tension and sub-glottal pressure.

With the integration of the 'plosive bellows' in his speaking machine, von Kempelen seems to have been far ahead of his time. It will remain the task of today's research to continue to develop models, that are able to temporally overcome the aerodynamic voicing constraint, an often-ignored phenomenon that was frequently pointed out by Ohala. Those models are important for demonstration and research, as was shown by Arai with "modern" material, e.g. in [2], or with replicas, e.g, [25].

Acknowledgements

The author would like to thank the editors, two anonymous reviewers and colleagues from Saarbrücken for their suggestions that substantially helped to improve the quality of this paper.

References

[1] ADJARIAN, H. Les explosives de l'ancien arménien étudiées dans les dialectes modernes. *La Parole ou Revue internationale de Rhinologie, Otologie, Laryngologie et Phonétique expérimentale* (Paris), pp. 119–127, 1899.

[2] ARAI, T.: Education in acoustics and speech science using vocal-tract models. *The Journal of the Acoustical Society of America* 131(3), pp. 2444–2454, 2012.

[3] BIRKHOLZ, P.: *VocalTractLab* [software] Available online at: http://www.vocaltractlab.de, 2017.

[4] BIRKHOLZ, P., STONE, S & KÜRBIS, S.: Comparison of different methods for the voiced excitation of physical vocal tract models. *Proc. Elektronische Sprachsignalverarbeitung (ESSV),* Dresden, pp. 84–94, 2019.

[5] BIRKHOLZ, P. & ZHANG, X.: Accounting for microprosody in modelling intonation. *Proc. IEEE International Conference on Acoustics, Speech and Signal Processing (ICASSP),* pp. 8099–8103, 2020.

[6] BONNEAU, A.: Realization of French voiced fricatives by German learners as a function of speaker level and prosodic boundaries. *Proc. 18th International Congress of Phonetic Sciences (ICPhS)*, Glasgow, 5 pp., 2015.

[7] BRAUN, A.: *Zum Merkmal "Fortis/Lenis". Phonologische Betrachtungen und instrumentalphonetische Untersuchungen an einem mittelhessischen Dialekt*. Stuttgart: Steiner, 1998.

[8] BRAUN, A.: An early case of "VOT". *Proc. Interspeech* 2013, Lyon, pp. 119–122, 2013.

[9] CATFORD, I.: *Fundamental Problems in Phonetics*. Edinburgh: Edinburgh University Press, 1977.

[10] CHO, T. & LADEFOGED, P.: Variation and universals in VOT: evidence from 18 languages. *Journal of Phonetics* 27, pp. 207–229, 1999.

[11] HOUSE, A. S. & FAIRBANKS, G.: The influence of consonant environment upon the secondary acoustical characteristics of vowels. *Journal of the Acoustical Society of America* 25, pp. 105–113, 1953.

[12] JESSEN, M.: *Phonetics and Phonology of Tense and Lax Obstruents in German*. Amsterdam: Benjamins, 1998.

[13] KEATING, P., LINKER, W. & HUFFMANN, M.: Patterns of allophone distribution for voiced

and voiceless stops. *Journal of Phonetics* 11, pp. 277–290, 1983.
[14] VON KEMPELEN, W. [1791]: *Mechanismus der menschlichen Sprache / The Mechanism of Human Speech.* Kommentierte Transliteration & Übertragung ins Englische / Commented Transliteration & Translation into English by FABIAN BRACKHANE, RICHARD SPROAT & JÜRGEN TROUVAIN (eds) Part 1 (Studientexte zur Sprachkommunikation Band 87) and Part 2 (Studientexte zur Sprachkommunikation Band 88). TUDpress: Dresden, 2017.
[15] KLEBER, F., JOHN, T. & HARRINGTON, J.: The implications for speech perception of incomplete neutralization of final devoicing in German. *Journal of Phonetics* 38, pp. 185–196, 2019.
[16] KRAEHENMANN, A. & LAHIRI, A.: Duration differences in the articulation and acoustics of Swiss German word-initial geminate and singleton stops. *Journal of the Acoustical Society of America* 123, pp. 4446–4455, 2008.
[17] LISKER, L. & ABRAMSON, A.S.: A cross-language study of voicing in initial stops: Acoustical measurements. *Word* 20, pp. 384–422, 1964.
[18] MADDIESON, I.: Voicing in plosives and fricatives. In: M.S. DRYER & M. HASPELMATH (eds.) *The World Atlas of Language Structures Online*. Leipzig: Max Planck Institute for Evolutionary Anthropology, 2013. (Available online at http://wals.info/chapter/4)
[19] MADDIESON, I.: Voicing and gaps in plosive systems. In: M.S. DRYER & M. HASPELMATH (eds.) *The World Atlas of Language Structures Online*. Leipzig: Max Planck Institute for Evolutionary Anthropology, 2013. (Available online at http://wals.info/chapter/5)
[20] OHALA, J.J. & RIORDAN, C.J: Passive vocal tract enlargement during voiced stops. In: J.J. WOLF & D. H. KLATT (eds.), *Speech communication papers*. New York: Acoustical Society of America, pp. 89–92, 1979.
[21] OHALA, J.J.: The origin of sound patterns in vocal tract constraints. In: P.F. MACNEILAGE (ed.), *The Production of Speech*. New York: Springer, pp. 189–216, 1983.
[22] OHALA, J.J.: Accommodation to the aerodynamic voicing constraint and its phonological relevance. *Proc.* 17th *International Congress of Phonetic Sciences*, Hong Kong, pp. 64–67, 2011.
[23] SVIRSKY, M. A., STEVENS, K. N., MATTHIES, M. L., MANZELLA, J., PERKELL, J. S., & WILHELMS-TRICARICO, R.: Tongue surface displacement during bilabial stops. *Journal of the Acoustical Society of America* 102, pp. 562–571, 1997.
[24] TANNER, J., SONDEREGGER, M., STUART-SMITH, J. & FRUEHWALD, J.: Toward "English" phonetics: Variability in the pre-consonantal voicing effect across English dialects and speakers. *Frontiers in Artificial Intelligence* 3, Article 38, 15 pp., 2020.
[25] TROUVAIN, J. & BRACKHANE, F.: Wolfgang von Kempelen's speaking machine as an instrument for demonstration and research. *Proc.* 17th *International Congress of Phonetic Sciences*, Hong Kong, pp. 164–167, 2011.
[26] WESTBURY, J. H.: Enlargement of the supraglottal cavity and its relation to stop consonant voicing. *Journal of the Acoustical Society of America* 73, pp. 1322–1336, 1983.
[27] ŻYGIS, M., FUCHS, S. & KOENIG, L.L.: Phonetic explanations for the infrequency of voiced sibilant affricates across languages. *Laboratory Phonology* 3(2), pp. 299–336, 2012.

Roots of Electronic Speech Communication in Linguistics and Engineering: the Genesis of the Chair of Speech Communication of the TU Dresden as a Case Study

Rüdiger Hoffmann, Dieter Mehnert
Technische Universität Dresden, Institut für Akustik und Sprachkommunikation
ruediger.hoffmann@tu-dresden.de

Abstract: Speech Communication is an interdisciplinary topic, but in the academic practice it has to be assigned to disciplines like linguistics and phonetics, electrical engineering, etc. In the case of engineering, the interest of the faculties originated in the first years mainly from scientific problems in telecommunication. But around the end of the 1960s, the focus widened to more general aspects of human-human and human-machine communication. The paper investigates how this process advanced at Dresden Technical University (TU) with special respect to external influences, originating from phonetics and communication sciences.

1 Motivation

Two years ago, in 2019, the Institute of Acoustics and Speech Communication of the TU Dresden celebrated its 50th anniversary [9]. When the Faculty of Electrical Engineering at TU Dresden was re-organized in the framework of the so-called *3rd higher education reform* of the German Democratic Republic (GDR) in 1968/69, a laboratory was installed, which aimed to address the future requirements, which were expected to occur at the interfaces of the computer with its environments as well as with the human user. The name was at first *Laboratory of Communication and Measuring*. This decision was a very early one, compared to other universities, and we count it as the founding act of the recent institute.

The laboratory included four working groups, among them one for *Speech Communication*, headed by Walter Tscheschner (1927–2004) as lecturer since 1969 and as full professor since 1972. Until his retirement in 1992, he supervised 37 Dr.-Ing. and Habilitation theses in the field of speech synthesis and recognition. The scientific results were applied in numerous projects with the industrial main partner *Kombinat Robotron* (computers) and others. He started many teaching activities in speech communication at the TU Dresden and outside. Although the Cold War strongly limited the interaction with the international community, the scientific level of his group remained high. Hence, the integration into the Western system posed no serious problems after the German re-unification, and the chair is working today continuously with the second successor of W. Tscheschner.

Details of this half-century success story have been published at different opportunities, e. g. [8]. But a kind of operational blindness prevented so far a discussion of the phenomenon that founding a chair for *speech communication* was more than could be expected from a faculty of electrical engineering as early as 1969. The "normal" requirements of such a department in that time might include chairs for *speech acoustics* or for *speech transmission*, dealing with the phenomenon of speech as physical signal.

The name of the Dresden group followed ideas of the phonetician Paul Menzerath (1883–1954) and his successor, the physicist Werner Meyer-Eppler (1913–1960), who introduced a unified view of phonetics, linguistics, and communication in the *Institute of Communication Research and Phonetics* at Bonn University [21]. When W. Tscheschner published his first teaching material after his appointment [26], Meyer-Eppler's communication model was a central paradigm.

This paper is devoted to the question how it happened that an engineering faculty installed a chair for speech communication so surprisingly early. Following this introduction, the paradigm shifts of the 1950s in linguistics and communication sciences are considered (Section 2). Then, the foundation of the *Institute of Phonetics and Communication Science* at Berlin Humboldt University is described briefly (Section 3). Section 4 depicts the early activities in speech analysis and transmission at the *Technische Hochschule* (TH) Dresden. In the main part (Section 5), it is shown how all these activities were (not only, but also) leading to the foundation of the Dresden chair and to many cooperations with other institutions. Section 6 concludes the paper.

2 Aspects of Linguistics in the Early Years of GDR

2.1 Wolfgang Steinitz and the Structural Grammar

Despite of the decline of the sciences in the years of the "Third Reich", some activities could be continued over the war time. E.g., the Prussian Academy of Sciences included a *German Commission* since 1903, which inter alia continued to edit the *German Dictionary* of the BROTHERS GRIMM. In 1946, the *German Academy of Sciences* was founded in Berlin in the Soviet occupation zone (since 1949 the GDR). It should not only continue the tradition of the Prussian Academy, but also include a network of research institutions, following the Soviet model.

In the framework of the Academy, the *German Commission* was upgraded to an *Institute for German Language and Literature* in 1952. As director, THEODOR FRINGS (1886–1968) was appointed, who was also the director of the *Germanistic Institute* of Leipzig University since 1927. His Academy institute in Berlin included a department for "Present German Language". There, the "Dictionary of the Present German Language" was elaborated, which is valued as "the most important lexicographic achievement in the German language of the 20th century" [17]. The department was headed by W. STEINITZ.

WOLFGANG STEINITZ (1905–1967, Fig. 1) was an exceptional scientist. His original field was Finno-Ugristics, but he is also known as an important folklorist. When he returned from Swedish exile to Berlin in 1946, he not only continued this work. He started broad activities in the organization of the scientific life in Eastern Germany and moved forward the development of Slavic studies and German linguistics.

The latter was necessary, because the new approaches, originated by FERDINAND DE SAUSSURE (1857–1913) and KARL BÜHLER (1879–1963), played no role hitherto in research and teaching. When in the 1950s the fundamental books of ROMAN JAKOBSON (1896–1982) and NOAM CHOMSKY (* 1928) appeared, a place was required in the GDR where linguistic innovations could be investigated and applied.

This place was organized by W. STEINITZ, who was a friend of ROMAN JAKOBSON since his exile in Sweden, at the German Academy of Sciences Berlin. The work started in his aforementioned department in 1956, but in 1959 he proposed the installation of a separate unit "for mathematical linguistics and translation machines" [1]. The implementation started in 1961 with the foundation of a commission with two working groups, which were upgraded to independent units in 1962. One of the units, the *Department of Structural Grammar*, played an important role in the "transition from the empiric to a theory-based linguistics", especially with respect to the generative grammar of N. CHOMSKY [7]. Until 1966, the department was directed by ALEKSANDR V. ISAČENKO (1910–1977), who developed a pioneering prosody model in partial cooperation with W. TSCHESCHNER in Dresden (cf. 5.3).

In his plan from 1959, W. STEINITZ proposed the personal inclusion of G. F. MEIER, at that time professor in Leipzig and then in Berlin. Indeed, he was installed as a member of the commission and as the head of one of the working groups. His work was very important for further development and will be sketched briefly in the following section.

Figure 1. WOLFGANG STEINITZ, photograph by ABRAHAM PISAREK, ca. 1948. Source: SLUB Dresden / Deutsche Fotothek / Abraham Pisarek (left).

Figure 2. GEORG F. MEIER, cutout from a private photograph from 1975 (right).

2.2 Georg F. Meier and the Communication Sciences

GEORG FRIEDRICH MEIER (1919–1992, Fig. 2) represented the generation who came in the horror of the war at young age, before they were able to start their studies. Gifted in foreign languages, he started to attend a broad program of studies in speech and language in Munich in 1946, which he continued in Halle (GDR) in 1949. He defended his PhD thesis in 1953 and his Habilitation thesis in 1958, both at Leipzig University.

G. F. MEIER was an outstanding scientist, governing a wide area from formal linguistics to phonetics and communication sciences. The Leipzig University (since 1953 named after KARL MARX) appointed him 1958 to the director of the *Institute of Indo-European Studies*. This was an institute with long tradition, including the roots of modern linguistics, which were created by H. G. C. VON DER GABELENTZ (1840–1893).

The institute was re-named *Institute of Linguistics* in 1959. G. F. MEIER explained the expanded profile in a letter from 1958 (following the citation in [6]): "The new tasks include, apart from the existing, comparative linguistics also in Non-Indo-European languages, research and teaching in the field of general linguistics, especially language theory, research in the field of Marxist linguistics as well as research and teaching in a number of special topics like phonetics, speech acoustics, problems of the second signal system, speech psychology, speech pathology, structural analysis, preliminary work at the problems of translation machines, etc."

Since that time, the appointment area of G. F. MEIER was *Linguistics and Communication Sciences* (with small variations). In Leipzig, however, he owned only the status of a guest professor, because he got a call at the Humboldt University in Berlin also in 1959.

The comparable task in Berlin was upgrading the existing *Institute of Indo-European Studies*. It was re-named *Institute of Linguistics* in 1959 and obtained a working group for *Linguistic Communication Research*, which was headed by G. F. MEIER, who immediately started extensive teaching activities in the study program "General and comparative linguistics and communication research".

The situation changed after the construction of the Berlin Wall in 1961. The head of the *Institute of Phonetics*, URSULA FEYER (1901–1989), remained in West Berlin, and the leaderless institute offered itself as a place for the implementation of G. F. MEIER's conception.

3 The Institute of Phonetics and Communication Sciences in Berlin

The new institute of the Humboldt University was founded on November 1, 1961, and received the name *Institute of Phonetics and Communication Sciences* (IPK). When its organization was completed in 1964, it included the following former units:

- *Institute of Phonetics*, founded in 1934 by the Africanist DIEDRICH WESTERMANN (1875–1956) as *Institute of Sound Research*. It was based on the "sound archives", which were collected since the beginning of the 20th century.
- *Institute of Rhetoric*, a smaller institute, which was guided in the end by HERTA RECLAM (1901–1988). It was partially adopted by the IPK in 1962.
- *Institute of Special Education*, which maintained the tradition of the speech therapists and special pedagogues HERMANN GUTZMANN sen. (1865–1922) and FRANZ WETHLO (1877–1960). This institute contributed to the IPK by the transfer of the former staff member GERHART LINDNER (1925–2005), who was appointed Professor for Phonetics at the IPK in 1964.
- *Laboratory of Experimental Phonetics and Acoustics*, also originating thanks to H. GUTZMANN and F. WETHLO, located at the ORL clinics of the Charité. This laboratory was integrated into the IPK with the appointment of G. LINDNER in 1964.

With this, the IPK incorporated a critical mass, based on the best traditions of the Berlin University in speech sciences [19]. G. F. MEIER developed the IPK to an important center of research and education. His teaching activities were re-organized in 1966 with the implementation of a course "Communication Science". Here is not the place to discuss all the scientific achievements. However, the high degree of networking should be mentioned with respect to the focus of this paper. An overview was given by G. F. MEIER in a memorandum in 1962 [16, p. 170]: "The focus and the coordination of the work is the task of the IPK of the Humboldt University Berlin, that cooperates in specific questions with the Institute of Telecommunication of the TU Dresden, the Institute of Machine Computing of the TU Dresden, the Institute of Linguistics of the Karl Marx University Leipzig, and the Institute of Psychology of the Humboldt University as well as institutes of the socialist foreign countries. [...] Cooperation exists also [...] with the working unit for mathematical linguistics and automatic translation of the German Academy of Sciences, which deals with the definition of algorithms for the translation English-German."

Communication sciences and linguistics formed only a part of the memorandum, which in toto aimed to the installation of a *Section for Cybernetics* at the Academy of Sciences, which was founded in November 1962. It included 21 members, among them G. F. MEIER and A. V. ISAČENKO from the field of linguistics and H. FRÜHAUF (radio engineering) and H. KINDLER (automatic control) from the Faculty of Electrical Engineering of TU Dresden. Cybernetics as interdisciplinary science offered improved opportunities for networking between social sciences, natural sciences, and engineering not only in general, but also at the institutional level.

4 Speech Analysis and Transmission at the TH Dresden before 1961

4.1 Introductory Remarks

With this section, the focus of the paper turns to engineering at the TH (Technische Hochschule) Dresden before 1961, the year of building the Berlin Wall. This year forms a caesura, because the political circumstances changed rapidly, which also influenced the scientific life. Furthermore, the TH Dresden was upgraded to a Technical University (TU) in this year, which is also important for our considerations, because W. TSCHESCHNER defended his Dr.-Ing. thesis.

Communication engineering was established in Dresden by HEINRICH BARKHAUSEN (1881–1956) with the designation "Schwachstromtechnik". This term was coined as the counterpart to "Starkstromtechnik" (power engineering) to give the emerging fields like electronics, telecommunication, and electroacoustics a common name. After the war destruction and the re-opening of the TH, BARKHAUSEN was active until his retirement in 1953. Several chairs were installed to cover the growing variety of tasks in this field. Among the heads of the new institutes, three professors are important for the early development of speech technology at TH/TU Dresden:

- HANS FRÜHAUF (1904–1991), founding director of the *Institute of Radio Engineering and Electron Valves* in 1950 and very engaged in different government bodies of the German Academy of Sciences,
- WALTER REICHARDT (1903–1985), founding director of the *Institute of Electroacoustics and Building Acoustics* in 1950,
- KURT FREITAG (1901–1977), founding director of the *Institute of Telecommunication* in 1954.

H. BARKHAUSEN was very interested in acoustics, but did not deal with specific problems of speech. First activities of the TH Dresden towards "visible speech" are verifiable at the renowned *Scientific-Photographic Institute* of HELLMUT FRIESER (1901–1988), who was engaged in the development of the sound film and adopted a method for spectral analysis of sound by optical-photographic means [2]. Between 1952–1957, the generation of "visible speech" by electronic means formed a project at the institute of W. REICHARDT [11]. But the actual impulse for the development of speech technology at the TH/TU Dresden came from K. FREITAG's *Institute of Telecommunication* with the development of a channel vocoder.

4.2 The Vocoder of the Institute of Telecommunication

With the end of World War II, the vocoder raised strong interest in different branches. There was an obvious question whether this innovation was really able to transfer speech economically with reasonable quality. The best way to find an answer was the investigation of a prototype, which had to be built following the information published so far.

At the TH Dresden, H. FRÜHAUF stressed the importance of the vocoder in 1952 [4]. It was decided to build an experimental system at the *Institute of Telecommunication* under the guidance of K. FREITAG. The channel vocoder of the British Post Office [5] served as a rough model. The implementation took place between 1954–1957 in the framework of the Dr.-Ing. thesis of EBERHARD KROCKER (* 1927) [15].

When E. KROCKER left the institute in 1959, remaining experiments with the vocoder were performed by W. TSCHESCHNER. Finally, it had to be assessed that the importance of the vocoder at that time was low for speech compression, but invaluably high for the investigation of the speech signal itself.

4.3 Early Years of Walter Tscheschner at the Institute of Telecommunication

W. TSCHESCHNER (Fig. 3) came in 1950 as a student of electrical engineering to the TH Dresden, the place, where he would turn into a highly renowned pioneer in electronic speech communication. Previously, his education had been interrupted by the war, he was wounded, and he was expelled from Silesia to Chemnitz, where he started his higher education, changing then to Dresden.

He chose the *Institute of Telecommunication*, because he expressed his interest in problems of transmission and remote control. There he came for the first time in touch with speech as electrical signal. It seems that he dealt with measuring the so-called reference attenuation already in a student project, before this became the task for his diploma thesis.

The reference attenuation is used in telecommunication for evaluating the quality of speech transmission of a telephony system by comparison with a reference system, the "calibration circle". The principle was known since ca. 1940, and K. FREITAG wanted to have such a

Figure 3. W. TSCHESCHNER in front of the Dresden vocoder. Newspaper photograph from 1962 [13].

system at his institute. W. TSCHESCHNER had to develop the electronic components, while the electromechanical interfaces to the system under test, which are known as "artificial ear and mouth", should come from other sources.

After finishing the thesis, the university hired W. TSCHESCHNER as scientific assistant from October 1, 1954. The quality assessment of speech transmission systems continued to be the working field, which should be improved by new solutions. A proposal, to replace the simple test signals (sweeps) by speech-like signals or synthetic speech, existed in the literature [14]. This connected the task with the work at the vocoder.

Thus, a plan for the dissertation from 1955 formulates a preliminary goal: "Speech analysis and investigation of the influences of a transmission system with the aim of an objective representation of speech comprehensibility." The vocoder was intended as an experimental platform. It turned out, however, that comprehensibility was less dependent on the spectral shape of the signal, which could be examined with the vocoder, than on the temporal behavior of the spectral components. Therefore a decision was made in 1957 to direct the focus of further investigation to the role of transients in the speech signal.

4.4 The Dr.-Ing. Thesis of Walter Tscheschner

The task, which appeared at this early instance, was no more and no less than finding the physical description of the speech signal which would enable to model the perception of the sounds and their imitation in synthesis and recognition. Today we know that this would be a mission for decades. Nevertheless, the thesis of W. TSCHESCHNER contributed a large amount of invaluable data by subjective and objective investigations.

The subjective tests used natural speech material which was changed in amplitude, duration, and bandwidth, and was evaluated by listeners. For the objective measurements, special equipment for time-frequency analysis was required, which was not available. W. TSCHESCHNER developed a spectral analyzer following the heterodyne principle (known as *Suchton* method), which guaranteed a high frequency selectivity. An additional device was provided for "time decomposition". It worked with a series of rotating magnetic heads, which allowed the periodic presentation of a 20 ms time window to the analyzer.

The thesis was completed in 1960 [22]; selected results were published at the 4th ICPhS in Helsinki [23].

5 The 1960s as the Period of Cooperation

5.1 Overview on the Activities of the 1960s

W. TSCHESCHNER defended his Dr.-Ing. thesis on April 24, 1961. The standard procedure would now have led him away from the university. It is mainly thanks to G. F. MEIER that he remained at the institute. This decision made sure that research in electronic speech communication was not just an episode in the history of the TU Dresden.

G. F. MEIER himself reports about the support for W. TSCHESCHNER after a visit in Dresden. On December 12, 1960, he sends W. TSCHESCHNER a letter, including the following: "After the inspection of your equipment on November 18, [1960], which was very impressive for me, I made an effort at the Academy at Prof. FRÜHAUF and at the State Secretariat for Higher Education that your work will not only be definitely continued, but also be extended by the practical goal of transferring spoken language into written one and vice versa." He recommends him to prepare for a career as a university professor.

Meanwhile, it was necessary to find a project for financing these goals. We know from publications that W. TSCHESCHNER had to deal temporarily with a project on welding equipment for the production of semiconductors. But the application for a project on the "physics of spoken language" was finally successful and ensured his phase of habilitation from 1962 to 1966. This time was extremely labor intensive and productive, which we can only describe as an incomplete overview.

5.2 The Habilitation Thesis of Walter Tscheschner

Apart from a detailed overview on the state of the art in physical modeling of the production of speech, the work is devoted to the features of natural sounds and their combinations which are relevant for their comprehensibility. So far, it is a continuation of the ideas of the Dr.-Ing. thesis (cf. 4.4). The contents are extended by a substantial amount of new material like measurements with the *Kay Sonagraph*, which was not available before. Also, his own results are compared to a comprehensive compilation of data from literature that had become available in the meantime. Remember that the fundamental monograph by GUNNAR FANT (1919–2009) was published in 1960 [3].

The main difference to the Dr.-Ing. thesis is the growing role of synthetic speech, although the title of the work is "Speech and its Recognition". That is a contradiction only on the surface, because speech synthesis was still serving as an experimental method to clarify the acoustic structure of speech sounds. The production of synthetic speech for communication purposes was at that time at most a question for the future.

In the last section of the thesis, W. TSCHESCHNER develops an outline for a speech synthesis device that was implemented in the later 1960s. It is roughly characterized by the synthesis part of a channel vocoder, which is controlled by a sophisticated arrangement ("Potential function control") for modeling the envelope of the formants. The very successful paradigm shift to the (more powerful) formant synthesizer was a matter of the 1970s and is therefore beyond the consideration of this paper.

5.3 Cooperation Between Dresden, Berlin, and Leipzig

The cooperation between the Institute of Telecommunication in Dresden and the IPK in Berlin was at first useful for W. TSCHESCHNER. He was supported from Berlin with speakers for a library of logatoms, and with listeners for hearing experiments. G. F. MEIER possessed a *Kay Sonagraph* at the IPK, which was a rare and expensive instrument that could be jointly used. It was borrowed to Dresden from 1961 to 1963.

G. F. MEIER also offered the opportunity to W. TSCHESCHNER to read lectures with the title "The Physics of Spoken Language" at the Humboldt University in the academic year 1963/64. This was important as a prerequisite for his intended habilitation.

On the other hand, the IPK benefited by the engineering background of the Dresden partner. This concerned especially the field of measuring pitch (more precisely, the fundamental frequency), which is an everlasting problem in speech technology. W. TSCHESCHNER developed a prototype [24] for the application at both institutions. This was the starting point for prosody research at the IPK, where DIETER MEHNERT (* 1935) came to the institute from the industry in 1963. He took care of the further development of the pitch measuring device [18]. Later on, he wrote his Dr.-Ing. thesis as well as his Habilitation thesis on aspects of speech prosody [19].

The availability of pitch measurement equipment led to further cooperations. Although the power of the devices was limited and the results were very preliminary, the projects are early attempts to build bridges between social sciences and engineering, thus representing the mind of this "cybernetic" decade. The following activities are known in more or less detail:

- *Applications in musicology*: We know about the exchange of ideas and preliminary experiments with the ethnomusicologist JÜRGEN ELSNER (* 1932, Musicological Institute of the Humboldt University) about soloist interpretations of songs, and ERICH STOCKMANN (1926–2003), a collaborator of W. STEINITZ at his Institute of German Folklore of the German Academy of Sciences, about the analysis of Albanian folk songs.
- *Applications in tonal languages*: The activities in Asian, African, and Latin America studies were concentrated at the Karl Marx University Leipzig in the 1960s. Research in tonal languages was a scientific focus. Two measuring campaigns were performed in 1964/65, one with the East Asian specialist EBERHART RICHTER (1935–1989) on samples from Burmese and Tibetan, and the second with the Africanist SIEGMUND BRAUNER (* 1934) on samples from Bambara, a language spoken in the Republic of Mali. Later on, these activities have been continued with D. MEHNERT at the Humboldt University.

Other equipment in Dresden like the vocoder and the heterodyne spectrometer were applied in further cooperations, as with the film industry (DEFA), and with ORL clinics. When A. V. ISAČENKO from the German Academy of Sciences elaborated his prosody model together with H. J. SCHÄDLICH (* 1935, later a writer of novels), he used the Dresden vocoder to generate synthetic intonation contours in 1962/63 [12].

5.4 Computerization and the Phase of the 3rd Higher Education Reform

The planning organs of the GDR responded to the growing importance of microelectronics and computing by measures in industry and science. The capacities for the development and production of computers were coordinated in a corporate association (*Vereinigung Volkseigener Betriebe*) in two steps (1961 and 1964), which was the predecessor of the *Kombinat Robotron*, founded in 1969 [20]. In the preceding units, aspects of human-computer interaction were already being considered. So it is not surprising that W. TSCHESCHNER changed to one of them, when he was in duty to work in the industry for a certain time according to the rules for future university professors.

To be specific, he entered the *VEB Elektronische Rechenmaschinen* in 1967 and worked in 1968/69 as project leader in the Main Research Center of *Robotron*, both in Karl-Marx-Stadt (now again Chemnitz). This phase has not yet been adequately investigated, but we know from two preserved reports that a research contract was signed between the Robotron Main Research Center, the TU Dresden, and the IPK of the Humboldt University.

When W. TSCHESCHNER returned to TU Dresden in 1969, the structures of the former Faculty of Electrical Engineering had been changed considerably. This was a result of the 3rd

Higher Education Reform of the GDR, which was intended to respond to new demands of the 1960s in society, science, and technology. The reform is seen controversially today, because it increased the influence of the state and party organs at the universities. But the engineering faculties benefited from the new structures, because they reflected the challenges of the upcoming electronic age. In this way, the *Laboratory for Communication and Measuring* was installed at the TU Dresden. It integrated the three institutes, which we mentioned in 4.1 (the acoustics institute in toto, the others partially), thus creating the space for new chairs, among them that of W. TSCHESCHNER. The further development was briefly sketched in Section 1.

6 Conclusion

This paper describes the genesis of a chair of Speech Communication at an engineering faculty in 1969, which was an early and rare event. The considerable influence of external institutions namely from linguistics and communication science was investigated here for the first time and was surprising even for the authors.

The paper is a preliminary result of a project which aims to publish a monograph about the history of speech technology research at the TH/TU Dresden. A first volume is planned to appear in 2022. It will among others include the material, which could be presented here in abridged form only, as complete as possible. A supplementary volume, containing the bibliography, was already published [10].

Acknowledgements

This work was enabled by availability of the scientific estate of WALTER TSCHESCHNER. The authors would like to thank his widow RUTH TSCHESCHNER and their children for generously giving numerous documents to the archive of the HAPS, including the correspondence with GEORG F. MEIER. We are especially grateful to BARBARA MEIER-ROEBER for supporting our project with supplemental information on the background of the cooperation of both scientists.

Special thanks also go to the reviewers and the editors for valuable hints.

References

[1] ABBAW (Archiv der Berlin-Brandenburgischen Akademie der Wissenschaften): AKL (1945–1968), Nr. 114; Schreiben von W. STEINITZ an Präsidium der DAW 29. 7. 1959.

[2] BECKER, H. E. R.: Die Analyse von Niederfrequenz durch Lichtbeugung an Kapillarwellen. *Annalen der Physik, 5. Folge, 36*, no. 7, pp. 585–608, 1939.

[3] FANT, G.: *Acoustic Theory of Speech Production*. 's-Gravenhage: Mouton & Co., 1960.

[4] FRÜHAUF, H.: *Moderne Verfahren der elektrischen Nachrichtenübertragung*. Berlin: Aufbau-Verlag, 1952.

[5] HALSEY, R. J.; SWAFFIELD, J.: Analysis-Synthesis Telephony with Special Reference to the Vocoder. *Journal of the IEE, Part I: General, 95*, no. 91, pp. 333–334; and *Part III: Radio and Communications Engineering, 95*, no. 37, pp. 391–406, 1948.

[6] HEHL, U. V.; JOHN, U.; RUDERSDORF, M. (eds.): *Geschichte der Universität Leipzig 1409–2009. Band 4: Fakultäten, Institute, Zentrale Einrichtungen*. Leipziger Universitätsverlag, 2009.

[7] HELBIG, G.: Zu Entwicklungen der germanistischen Grammatikforschung in der ehemaligen DDR (ein Rückblick). *Deutsch als Fremdsprache, 28*, no. 2, pp. 67–76, 1991.

[8] HOFFMANN, R.: Sprachsynthese an der TU Dresden: Wurzeln und Entwicklung. In: WOLF, D. (ed.): *Beiträge zur Geschichte und neueren Entwicklung der Sprachakustik und Informationsverarbeitung. Werner Endres zum 90. Geburtstag*. Dresden: Universitätsverlag, pp. 55–77, 2005.

[9] HOFFMANN, R.: 50 years Institute of Acoustics and Speech Communication; 30 years Conference Electronic Speech Signal Processing; 20 years Historic Acoustic-Phonetic Collection. In: BIRKHOLZ, P.; STONE, S. (eds.): *Elektronische Sprachsignalverarbeitung 2019. Tagungsband der 30. Konferenz*. Dresden, pp. 1–8, 2019.

[10] HOFFMANN, R.: *Bibliografie zur Geschichte der Sprachtechnologie an der Technischen Universität Dresden bis 2019*. Dresden: TUDpress, 2020.

[11] HOFFMANN, R.: Der historische Beitrag der TH/TU Dresden zur Entwicklung von Sprachsichtgeräten. In: *DAGA 2021, 47. Jahrestagung für Akustik*. Vienna, August 15–18, 2021.

[12] ISAČENKO, A. V.; SCHÄDLICH, H.-J.: Untersuchungen über die deutsche Satzintonation. In: *Untersuchungen über Akzent und Intonation im Deutschen*. Berlin: Akademie-Verlag (Studia Grammatica, vol. VII), pp. 7–67, 1965.

[13] K[LEFFE], H.: Die Geisterstimme aus Dresden. *Wochenpost*, no. 36, p. 11, 1962.

[14] KOSCHEL, H.: Künstliche Sprache. *Feinwerktechnik, 54*, no. 12, pp. 313–316, 1950.

[15] KROCKER, E.: Aufbau und Untersuchung eines Übertragungssystems für synthetische Sprache. *Wissenschaftl. Ztschr. der TH Dresden, 6*, no. 4, pp. 757–776, 1956/57.

[16] LIEBSCHER, H.: *Fremd- oder Selbstregulation? Systemisches Denken in der DDR zwischen Wissenschaft und Ideologie*. Münster: Lit (Selbstorganisation sozialer Prozesse, vol. 2), 1995.

[17] MALIGE-KLAPPENBACH, H.: *Das "Wörterbuch der deutschen Gegenwartssprache"*. Ed. by FRANZ JOSEF HAUSMANN. Tübingen: Niemeyer (Lexicographica, Series Maior, vol. 12), 1986.

[18] MEHNERT, D.: Einrichtungen zur Mehrfachanalyse akustischer Signale. *Proc. 6th Int. Congress of Phonetic Sciences, Prague, September 9–13, 1967*. Prague: Academia Publishing House, pp. 633–635, 1970.

[19] MEHNERT, D.: Phonetik an der Berliner Universität – ein Rückblick auf ihre Geschichte und auf Forschungsarbeiten der letzten Jahre. In: WOLF, D. (ed.): *Beiträge zur Geschichte und neueren Entwicklung der Sprachakustik und Informationsverarbeitung. Werner Endres zum 90. Geburtstag*. Dresden: Universitätsverlag, pp. 33–54, 2005.

[20] MERKEL, G.: *VEB Kombinat Robotron, Sitz Dresden – Ein Kombinat des Ministeriums für Elektrotechnik und Elektronik der DDR*. Fassung vom 30. 12. 2005, URL: http://robotron.foerderverein-tsd.de/111/robotron111a.pdf

[21] MEYER-EPPLER, E.: Paul Menzerath 70 Jahre. *Zeitschrift für Phonetik und allgemeine Sprachwissenschaft, 7*, no. 1/2, pp. 146–149, 1953.

[22] TSCHESCHNER, W.: *Analyse der deutschen Sprache unter besonderer Berücksichtigung der nichtstationären Vorgänge*. Dissertation, TH Dresden, 1961.

[23] TSCHESCHNER, W.: Verfahren und Ergebnisse der Analyse deutscher Vokale und Konsonanten. *Proc. 4th International Congress of Phonetic Sciences, Helsinki, September 4–9, 1961*. The Hague: Mouton, pp. 280–298, 1962.

[24] TSCHESCHNER, W.: Ein Beitrag zur Messung der Stimmbandgrundfrequenz. *Proc. 3rd Acoust. Conf., Budapest, June 1–8, 1964*, pp. 456–461, 1964.

[25] TSCHESCHNER, W.: *Die Sprache und ihre Erkennung*. Habilitationsschrift, TU Dresden 1967.

[26] TSCHESCHNER, W.: *Grundlagen der Kommunikationstechnik*. 1. Lehrbrief. Berlin: VEB Verlag Technik 1972.

The information about the structural and personal development of the different institutions was extracted from the Yearbooks of the German Academy of Sciences and from the Semester Catalogues of the universities, resp.

Archive material, which is not explicitely cited due to the limited space, is located in the Historic Acoustic-Phonetic Collection (HAPS) of the TU Dresden.

LACERDA'S CHROMOGRAPHS (1930s–1950s): THE CIRCULATION AND APPROPRIATION OF KNOWLEDGE IN EUROPE AND THE AMERICAS

Quintino Lopes[1], *George Brock-Nannestad*[2]
[1]*Institute of Contemporary History – NOVA University of Lisbon; University of Évora, Portugal*
[2]*Copenhagen, Denmark*
quintinolopes1@gmail.com

Abstract: From the 1930s to the early 1950s, chromography, a technique invented and developed by Armando de Lacerda, constituted an advanced method of investigation in the field of phonetics which overcame the limitations of kymography, the method predominantly used in experimental phonetics laboratories during the period. The new technique was first used by Lacerda in collaboration with Paul Menzerath in Bonn, its use later spreading to Portugal and Brazil. The existence of the most advanced chromographic equipment at the University of Coimbra Experimental Phonetics Laboratory since 1936 explains why Portugal became a global centre for Phonetics research and remained such up until the 1950s, attracting numerous scientists from abroad who, under the supervision of its head, Armando de Lacerda, received specialist training in the use of chromography. Later, the technique was used at the University of Bahia, where Lacerda and Nelson Rossi established the first experimental phonetics laboratory in South America in 1956–57, and where the first work on linguistic geography of Brazil, entitled *Atlas Prévio dos Falares Baianos*, was produced.

Keywords: History of experimental phonetics; chromography; Armando de Lacerda; circulation of knowledge

1. Introduction

The aim of this chapter is to provide a contribution to increasing our knowledge about the development of experimental phonetics. In contrast with the field of linguistics as a whole, which has a long historiographical tradition, there are comparatively few studies focusing on the development and importance of the techniques of experimental phonetics, theories and instruments developed in the field, and the institutions and actors involved [1].

We firstly provide a background to the history of instruments used in the field of phonetics before focusing on chromographs. We reconstruct the path followed by the instruments invented and developed in the 1930s by Portuguese phonetician Armando de Lacerda, and examine the following questions: where these instruments were first used; for what purpose; what the equipment consisted of; how it worked; where was knowledge about its existence circulated; who appropriated such knowledge; what its impact was on the development of experimental phonetics in the first half of the 20th century; how far this led to the international circulation of scientists.

In geographical terms, our focus is on three research facilities equipped with chromographs: the University of Bonn Phonetics Institute, and the Experimental Phonetics laboratories of the universities of Coimbra in Portugal and São Salvador da Bahia in Brazil. By analysing links and exchanges between the three institutions, we seek to revive the memory in the field of historiography of the important role played by chromography and the relationship it established between researchers from a range of countries in different continents.

2. The History of Experimental Phonetics Instruments: some considerations

Experimental phonetics is the term used for describing the use of equipment or apparatus with a view to enhancing and quantifying the sensory perception of spoken language, and an impression of its history is required in order to understand the contribution of individual researchers in the field.

The development of recording or registering instruments is in itself fascinating and has been dealt with in depth by V. J. Phillips in his book *Waveforms* [2]; however, their specific application in the field of phonetics has not been covered. Three major accounts of the history of experimental phonetics are available: two books based on a collection of relevant equipment at the University of Dresden [3, 4], the proceedings of a conference [5], and Michael Ashby's dissertation on experimental phonetics at University College London [6], which covers its development in the UK. Furthermore, the linguist and phonetician Göran Hammarström has made an unpublished paper available to the authors, because it was considered important to have the views of a professional in this particular field, rather than the authors' look-in from the outside.

There follows a transcription of much of this unpublished paper:

> «What can be considered as a first period of experimental, also and better called instrumental, phonetics begins with Jean Rousselot establishing a phonetics laboratory in 1889 in Paris [5: 91-100]. This innovation was adopted by many European universities, but only by some universities in other parts of the world. Up to about 1950 the speech sounds were mainly investigated by kymography and palatography. Both were cumbersome techniques, and the results were not of the highest order. The kymograph showed the wave as a line traced by a stylus on sooted paper. The stylus also showed the airflow to some extent. It jumped up at the plosive phase of stop consonants. Most work concerned the duration of vowels and consonants. There was little regard for the difficulty of delimiting sounds that are co-articulated or that change gradually one into the other, but the results are on the whole valid. To make the stylus produce a good trace, or a trace at all, one had to say the sounds into the funnel held in front of the mouth in an unnaturally loud and clear way, which somewhat limits the adequacy of the results. It was normal that one looked at the just made kymogram to judge its quality. If it did not look good, it was discarded. Unfortunately, a good kymogram could then be discarded if the picture did not correspond to the phonetician's idea of the sound.
>
> In order to avoid the problem of the friction of the stylus against the sooted paper, which could influence the picture of the tracing, Lacerda constructed in the 1930s in Coimbra an instrument he called chromograph that used an ink jet on white paper [7]. The same procedure was later used in Siemens – Elema's mingograph. Another technically advanced instrument was the (cathode-ray) oscilloscope, which as oscillograph could show the sound wave on paper. Like the mingograph these instruments were known but not as much used as the kymograph by the early phoneticians.
>
> E.A. Meyer constructed a device for measuring on the kymograms what was, and still is, wrongly called pitch. It could be used in studies of intonation. The device produced a curve in Hertz (which refers to fundamental frequency, not to pitch). Lacerda constructed a similar device that he called the tonometric triangle to be used on the tracings of his chromograph [7].

Palatography was the most used and best-known technique to study articulation. One obtained pictures of where the tongue touches the palate in two ways. Either the palate was sprayed with carbon dust and the tongue removed it where it touched the palate, or the tongue was painted with black liquid and made an imprint on the palate. One could use a mirror and a camera to obtain a picture of the imprint. Another technique was to construct an artificial palate of thin flexible metal, which was introduced into the mouth and on which the imprint was made. It was then taken out and the imprint was drawn. The form of the lips could be easily photographed. An x-ray picture could be taken of the typical articulatory position of a vowel or a consonant, or an x-ray film could be taken of the articulation of segments of speech. The use of these important possibilities was limited probably because the phonetics laboratory did not have the equipment and because x-rays are dangerous.

One can consider that the second period of phonetics begins about 1950 because new instruments came into use. This happened immediately in some laboratories and more slowly in others. Excellent sound could be recorded by magnetophones, which later were just called tape recorders. The rather primitive cylindrical Edison phonograph had had some use in the earliest laboratories. The gramophone with disk and turntable was available since the establishment of practically all laboratories. Its sound was good but the recording was complicated and the equipment was not available in the laboratory so it was not used. New electronic pitchmeters and intensitymeters could show the curve of the fundamental frequency in Hertz and the intensity curve in decibels respectively.

The really revolutionary instrument was, however, the sonagraph [sound spectrograph], which was available from about 1950. It was now possible to analyse the acoustic picture of what one pronounced and heard as vowels, consonants, coarticulations and suprasegmentals. As one well knew the correspondences between acoustic and articulatory details, the acoustic results were at the same time articulatory results. To use the new instruments, one needed a technician. The sonagraph in particular required much attention to function perfectly. The technician also needed to be in attendance when the students used the instruments in their work. Less progressive phoneticians continued to work with the kymograph till at least towards the end of the 1950s» [8].

Thus, the equipping of the Coimbra Phonetics Laboratory with chromographs, documented from its foundation in 1936, coincided with a new stage in the history of experimental phonetics in the first half of the 20th century, and it is important that this should be recognised.

3. Chromography: from the local context to the global

Founded in 1936 and directed by Armando de Lacerda, the University of Coimbra Experimental Phonetics Laboratory was regarded in the early fifties by several members of the international academic community as the most advanced experimental phonetics laboratory in Europe. There is evidence of its excellent reputation, based on the high standard of the facilities at the laboratory, in the writing of Pierre Van Bever, for example, who published the following comment in *Revue belge de philologie et d'histoire* in 1953: "Le prof. Lacerda est lui-même l'inventeur de nombreux appareils de mesure et d'enregistrement, et son laboratoire, fondé en 1936, est sans doute le meilleur d'Europe" [9].

At the same time, in a similar vein, although not so peremptorily, Francis Millet Rogers (Harvard University) stated:

«In the Faculty of Letters of the University of Coimbra is a superbly installed and very well-equipped Laboratory of Experimental Phonetics. [...] Great credit is due the Portuguese Ministry of National Education, the Instituto de Alta Cultura, and administrative officials and professors of the University not only for having supported the founding of the Laboratory in 1936 but also for having so generously provided excellent space for it in the recently constructed new building of the Faculty of Letters. Credit is also due them, and the Laboratory's genial Director, Dr. Armando de Lacerda» [10].

Figure 1. Two of the ten rooms that the Coimbra Phonetics Laboratory occupied in 1952 in the new Faculty of Letters building, with "an endless paper strip emerging from a horizontal projection chromographic recorder" in the foreground. (Biblioteca Nacional de Portugal)

In 1979, about a quarter of a century after this photographic record was produced and the opinions above recorded, in the absence of a successor to Lacerda as head of the facility, the Coimbra Phonetics Laboratory was closed down. The same year, as the memory of the laboratory began to fade, its first and only head retained the distinction of honorary member of the Permanent Council for the Organization of International Congresses of Phonetic Sciences, along with Roman Jakobson (Cambridge, USA) and Eberhard Zwirner (Cologne, Federal Republic of Germany).

This distinction was bestowed personally on Lacerda at the 4th International Congress of Phonetic Sciences, held in Helsinki in 1961, during the course of a career launched during Germany's Weimar Republic, referred to by his disciple Göran Hammarström, founder of the University of Uppsala Department of Phonetics (1955), in the following terms:

«When the young Lacerda decided to study abroad, he went to Hamburg where the highly reputed Giulio Panconcelli-Calzia was working. There he constructed instruments to measure articulatory movements but, as he told me, Panconcelli-Calzia did not give him the praise he had hoped for. Therefore, he left for Bonn where Menzerath was the head of a phonetics department that would overtake Hamburg as the most important place to study phonetics in Germany. Menzerath appreciated Lacerda's work and suggested that they should write a book together. According to what Lacerda told me he wondered if this was a good idea because the suggested topic, coarticulation, had already been treated by other authors. But Menzerath had no hesitations so one of the most important books in phonetics was written:

Koartikulation, Steuerung und Lautabgrenzung (1933). It would be obvious that the elegant formulations and the general picture is Menzerath's, but Lacerda's basic ideas and results are crucial. The order of the names of the authors of the book, Menzerath – Lacerda, is the one which one expects when an older professor has collaborated with a younger disciple. As things are, one sometimes sees authors quoting Menzerath's ideas on coarticulation. This is not acceptable because Lacerda is perhaps the more important author» [11: 90].

In addition to his pioneering studies on the co-articulation and segmentation of speech sounds, as mentioned above, Armando de Lacerda also stands out for his creativity in inventing and developing new instruments. Among the various devices he designed during his time in Hamburg and Bonn is the polychromograph.

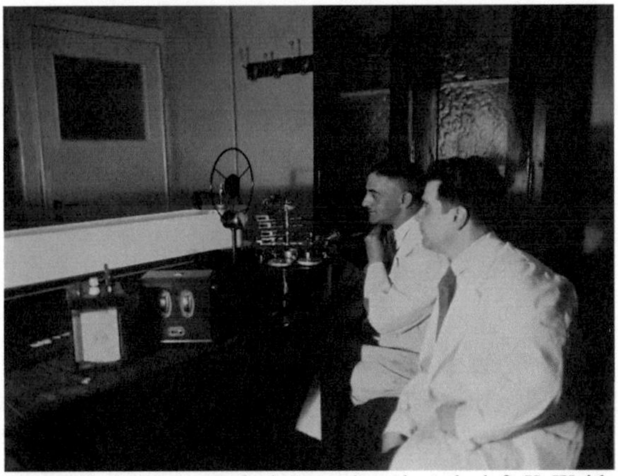

Figure 2. The photograph shows Armando de Lacerda and, on the left, K. Weitkuss, a laboratory assistant at the University of Bonn Institute of Phonetics, operating a polychromograph (1932). (Arquivo do Camões, Instituto da Cooperação e da Língua, I.P., Caixa 1337, Processo 3)

The polychromograph had a mouthpiece and recorded speech sounds and mouth movements on a strip of paper with a fine jet of ink, thus resolving the problem of stylus friction with the kymograph. Lacerda developed three methods of deflecting this jet. One was direct vibration from the *tambour registrateur* – this avoided all friction and the influence of the weight of the lever; another was sideways blowing on the jet (the most direct influence without electronic amplification), and the third was electromagnetic vibration of the nozzle. The electromagnetic solution meant that amplifiers were involved.

The development of the Lacerda polychromograph marked the introduction of chromography as a new research method and the inventor received funding of 11,000 escudos from the Portuguese state for its construction, provided that it was to be used at an experimental phonetics laboratory to be established in Portugal.

Before this happened, chromography was initially used in Germany. In 1932, Armando de Lacerda was invited by Eberhard Zwirner to use chromography with the "Ketterer system" at the Kaiser-Wilhelm Institut für Hirnforschung in Berlin. As early as 1936, signalling the growing international acceptance of chromography, Lacerda noted: "In a new paper in Dutch, Professor Theodor Baader from Nijmegen cites my chromographic method as one of the greatest advances in experimental phonetics" [12].

Recognising the importance of the new method, in 1933 Paul Menzerath invited the Portuguese phonetician to teach an 'Introduction to Chromography' course at the Institute of Phonetics at the University of Bonn.

Figure 3. The 'Introduction to Chromography' course taught by Lacerda was targeted at professors and PhD students of the University of Bonn. (Zamora Canellada Family Archive)

The course led to the use of chromography in PhD studies at the University of Bonn applied to phenomena of tonal variation in the Chinese and English languages. An example of this in the case of Chinese was the PhD dissertation of Schu-Yün Chwang-Wang (Bonn, 1935), whose phonograms were recorded with the use of chromography by Armando de Lacerda himself.

Returning to Portugal in July 1933, Lacerda set up the first Portuguese experimental phonetics laboratory at the University of Coimbra [13]. The laboratory opened in 1936, and its head stated that it would boast the most advanced chromographs, which explains the attraction it held for many scientists from abroad from the 1930s to the 1950s, including the likes of Peter Strevens (then head of the Department of Phonetics of University College of the Gold Coast), Francis Millet Rogers (Harvard University), Göran Hammarström (University of Uppsala), J. H. Uldall (University of Edinburgh) and Nelson Rossi (University of San Salvador da Bahia), among many others.

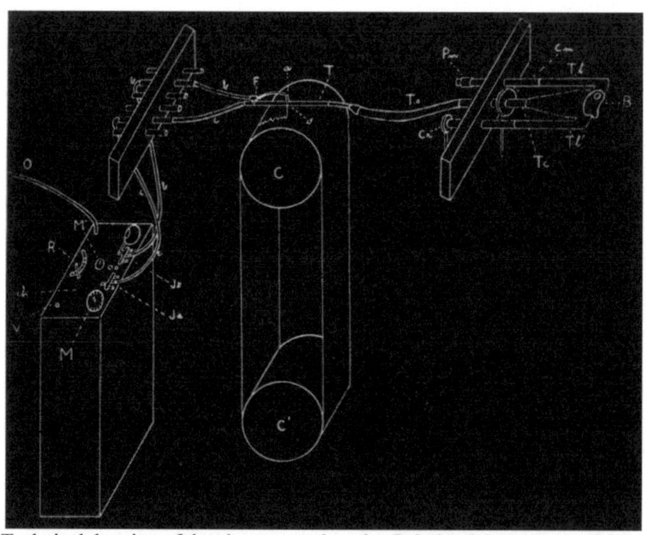

Figure 4. Technical drawing of the chromograph at the Coimbra laboratory in 1939. (Biblioteca do Laboratório de Fonética. Faculdade de Letras de Coimbra)

The main components of the chromograph can be seen in the drawing in Figure 4:

1) Sound recording devices, with the mouthpiece on the right (B). Made of metal plate, the mouthpiece was thick enough not to vibrate and much bigger than that used in kymography. The front part is oval in shape and the rear is conical, while the outside edges of the front part are bevelled in order to enable a good fit on the face. The space at the front is connected to two lateral pieces of tubing (Tl and Tl'), the rear extending into the central tube (Tc).

2) Chromographic projectors. Speech sounds are recorded by injecting coloured ink onto a paper strip running below. They are connected to the central tube (Tc) by means of a tube (Ts) and are fed ink, at the other end, through a rubber tube (b), next to the tube which feeds in air (c).

3) Pressure reservoirs and regulators. Feeding tubes (b and c) communicate with two sets of taps (Jt and Ja), for feeding ink and air, fixed on a metal plate (ch), which rests on a support table (compare Figure 6). Below the metal plate there are two reservoirs. A regulator (R) controls a tap which coordinates the flow from these reservoirs. A manometer (M) indicates air pressure in the ink reservoir and another manometer (M') indicates air pressure in the other (compensatory) reservoir, which is connected to a bottle of compressed air by means of a tube (O) (compare Figure 5).

4) Recording cylinder (C) powered by an a/c motor.

5) Auxiliary cylinder (C'), whose distance from recording cylinder can be regulated, allowing for the use of strips of paper [19: 8-18].

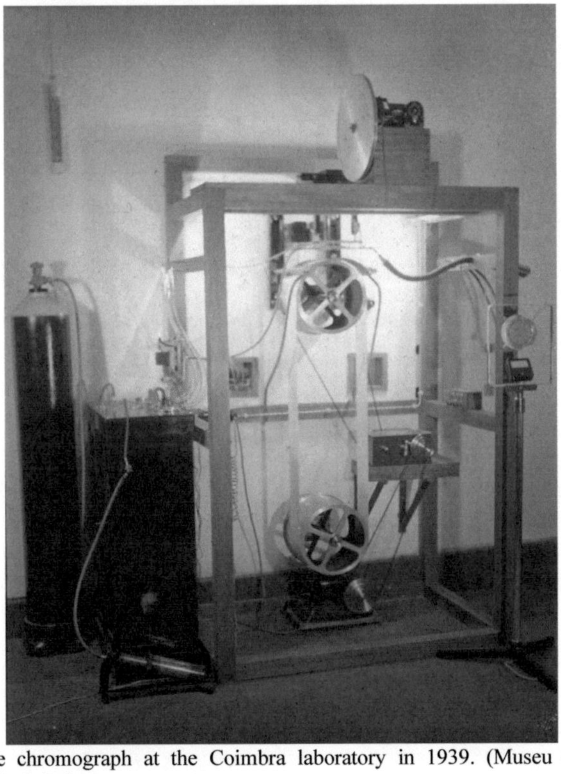

Figure 5. The chromograph at the Coimbra laboratory in 1939. (Museu da Ciência da Universidade de Coimbra)

Demonstrating how chromographs were decisive for many linguists electing Coimbra as their choice of university for receiving specialist phonetics training in the first half of the 20th century, there is a letter that Lacerda sent to Eberhard Zwirner in 1940. Regarding the work he published jointly with Francis M. Rogers as a result of their collaboration at the Coimbra Phonetics Laboratory in 1939, the Portuguese phonetician noted in the letter:

> «The work [...] was made with the collaboration of Mr. Francis Millet Rogers, [...] of the Graduate School of Arts and Sciences, Harvard University, who has been studying with me this year in my laboratory in order to obtain a first-hand knowledge of the latest developments in my chromographic method» [14].

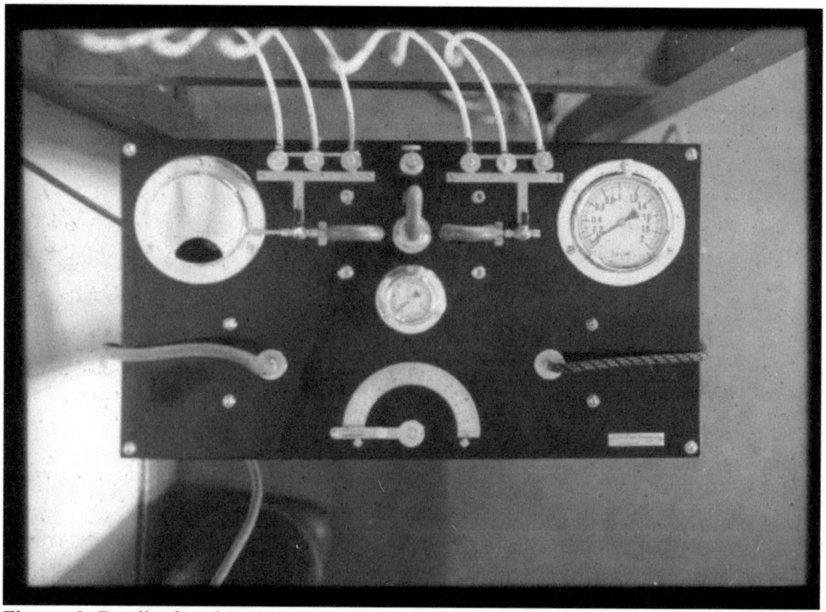

Figure 6. Detail of a chromograph used by Francis M. Rogers at the Coimbra Phonetics Laboratory, where he underwent specialist training for four months in 1939. (Museu da Ciência da Universidade de Coimbra)

On the impact on scientists' academic careers as a result of using chromographs in their work, we have the response given by Georges Straka, head of the Institute of Phonetics at the University of Strasbourg, to the head of the Uppsala University Faculty of Letters when asked about the scientific skills of Göran Hammarström: "His research on the length of phonemes in Swedish [...] is of great general interest to those in the field of linguistics. The results achieved far exceed those of previous researchers, thanks to the excellent chromographic method used by Mr. Hammarström" [15]. When one considers that this response in 1957 was decisive with regard to the advancement of Hammarström's academic career – he began as a PhD student at the Coimbra Phonetics Laboratory, later becoming the first full professor of Linguistics in Australia, at Monash University – the importance of conducting research using the chromograph at Coimbra becomes clear.

In the context of the development of the chromograph by Lacerda, Hammarström published a work in 1952 entitled *Le chromographe et le triangle tonométrique de Lacerda* [16]. Its publication, in French, illustrated with technical drawings, led to several requests from laboratories abroad for chromographic equipment. In this regard, what happened in Spain in the post-World War II period should be mentioned: after the failure of the Consejo Superior de Investigaciones Científicas – Higher Council for Scientific Research – to establish a phonetics laboratory at the University of Barcelona, to be equipped with Armando de Lacerda's instruments, in 1958 Manuel Companys, who had previously carried out research under the supervision of Lacerda at Coimbra, reaffirmed his intention to establish the use of chromography at a phonetics laboratory to be set up in Barcelona:

> «[...] it seems that we will be able to set up a laboratory in Barcelona next year, for which I will be responsible. Unfortunately, very little money will be available for the project. I plan to set up units dedicated to the following: HiFi recording [...]; spectrography [...]; chromography; recording and transcription [...]; surveys [...]; photography and macrophotography [...]. I would also like a speech-stretcher [...]. Unfortunately, the spectrographic equipment and probably the HiFi equipment [...] has to be dispensed with, but the last thing I will give up is our chromographic equipment, which we can do a lot with» [17].

In the 1950s, on the other side of the Atlantic, another university achieved what apparently did not come to fruition in Spain. As part of a concerted policy for the development of experimental phonetics in Brazil, Nelson Rossi of the University of São Salvador da Bahia obtained funding for undergoing specialist training at the Coimbra Phonetics Laboratory in 1954–55. After completing his internship working under Lacerda's supervision, he later became an intern at the Institut de Phonétique in Paris. Rossi then returned to São Salvador and the Portuguese phonetician was invited by the rector of the University of Bahia, Edgard Santos, to come to Bahia and set up what would be the first experimental phonetics laboratory in South America [7].

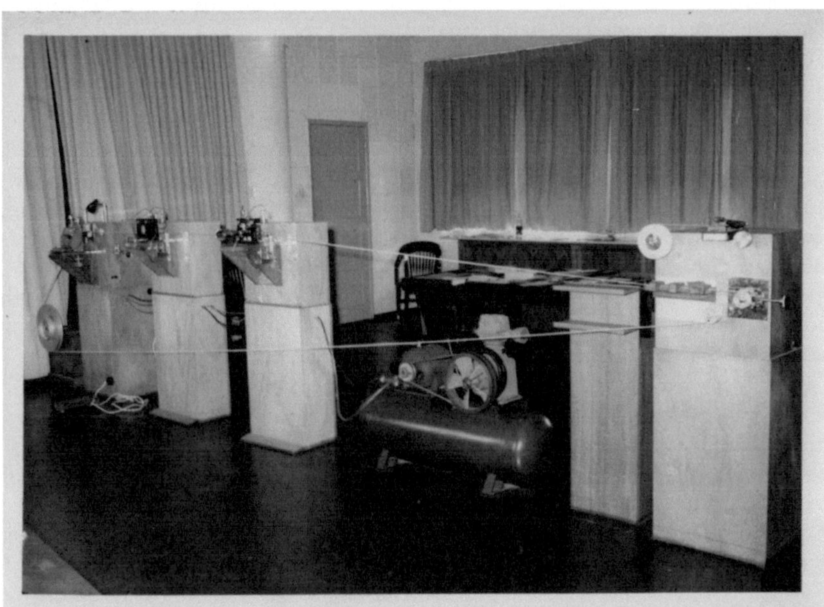

Figure 7. Equipped with chromographs, the University of Bahia Experimental Phonetics Laboratory carried out measurements and analyses, one of its main roles, using the techniques of chromography and tonometry, based on the model used at the Coimbra Phonetics Laboratory (Arquivo Familiar Paulo de Lacerda – Paulo de Lacerda Family Archive)

Highlighting the importance of the role of Lacerda and the research techniques he developed, associated with the establishment and operations of the Brazilian laboratory, Nelson Rossi, its head, stated:

> «Our aim is that the Phonetics Laboratory of the University of Bahia Faculty of Philosophy should work as closely as possible with similar bodies all over the world, and especially Coimbra, to exchange knowledge and ideas, not only due to the fact that the work of the two laboratories focuses on what is basically the same language [...] but above all because the Bahia laboratory was inspired by and uses the models and methods devised by the true founder of both institutions: Armando de Lacerda» [18: 6].

While Armando de Lacerda also set up another experimental phonetics laboratory at Colégio Pedro II in Rio de Janeiro in 1957–58, which was also equipped with his chromographic instruments, the laboratory in Bahia achieved greater standing in the field of linguistics and phonetics in Brazil. For example, it was the latter that produced, from 1960 to 1963, the first linguistic atlas of Brazil, entitled *Atlas Prévio dos Falares Baianos,* produced by Rossi and his collaborators and based on techniques of phonetic transcription appropriated from Armando de Lacerda and Göran Hammarström [7].

4. Conclusion

The history of scientific instruments shows that up until the early 1950s the kymograph and the palatograph constituted basic items of equipment at experimental phonetics laboratories around the world. From then on, the most advanced research facilities in this field of linguistics were equipped with the sound spectrograph. We seek to add a new chapter in the history of experimental phonetics by highlighting the importance of the polychromograph, invented by Armando de Lacerda in Bonn in 1932. Later, in Portugal, Lacerda used only one channel (for the speech waveform) and termed his instrument simply chromograph.

These instruments were not as much used as the kymograph. As the cost of production of instruments involving sound amplification fell, interest in the chromograph waned, and one may speculate that if it had been invented 10 years earlier it might have become standard equipment, replacing the kymograph and several types of amplified instruments until the cost of the sound spectrograph fell to a level which enabled it to be purchased by phonetics laboratories.

Despite this, until the appearance and dissemination of the sound spectrograph, the technical advance that the chromograph represented in comparison with the kymograph meant that laboratories that were equipped with the former boasted better facilities. Thus, in terms of history of experimental phonetics, and the history of science in general, it is interesting to note that Portugal, as it possessed an experimental phonetics laboratory equipped with the most advanced chromographs, had from 1936 to the early 1950s what came to be regarded by a number of influential linguists of the time as the best facility of its kind in Europe.

Some of the other laboratories in Europe and South America either sought to acquire chromographs or were equipped with such instruments during the period from the 1930s to the 1950s. Among these were the Experimental Phonetics Laboratory of the University of São Salvador da Bahia, the first laboratory of its kind in South America, whose work further demonstrates how the use of chromography spread outside Europe.

5. Acknowledgements

In 2019, following the 3rd International Workshop on the History of Speech Communication Research held in Vienna, I was contacted by the late George Brock-

Nannestad, who showed interest in studying the instruments that Armando de Lacerda devised for experimental phonetics. We soon decided to jointly publish a paper on the subject. Unfortunately, he passed away in December 2020 before this work was completed. As a consequence of this irreparable loss, any errors are the first author's responsibility. I am grateful for the privilege of having been able to collaborate with George Brock-Nannestad, and we also benefitted from the help and kindness of the late Göran Hammarström, who made available an unpublished paper of his. I am therefore grateful to Hammarström and would like to thank his daughter, Marika Hammarström, for giving her permission for it to be cited in the present chapter following the death of her father.

6. Funding

The Instituto de História Contemporânea (Institute of Contemporary History, University of Évora Unit) receives national funding from the FCT – Fundação para a Ciência e a Tecnologia under projects UIDB/04209/2020 and UIDP/04209/2020. This research was also supported by the private company Ferraz de Lacerda, Lda.

7. References

[1] KOERNER, E. F. K.: Historiography of Phonetics: The State of the Art. *Journal of the International Phonetic Association*, 23 (1), pp. 1–12, 1993.

[2] PHILLIPS, V. J.: *Waveforms: A History of Early Oscillography*. Bristol: IOP Publishing Ltd. Adam Hilger, 1987.

[3] MEHNERT, D.: *Historische phonetische Geräte, Katalog der historischen akustisch-phonetischen Sammlung (HAPS) der TU Dresden, erster Teil*. Dresden: TUD Press, 2012.

[4] MEHNERT, D., PÉTURSSON, M. & HOFFMANN, R.: *Experimentalphonetik in Europa. Institute, Wissenschaftler, Leistungen Ende des 19./Anfang des 20. Jahrhunderts*. Dresden: TUD Press, 2016.

[5] BOË, L.-J. & VILLAIN, C.-E. (Eds.): *Un siècle de phonétique expérimentale. Fondation et éléments de développement*. Lyon: ENS Éditions, 2010.

[6] ASHBY, M. G.: *Experimental phonetics in Britain, 1890–1940*. Unpublished doctoral dissertation, University of Oxford, 2016. Available at https://ora.ox.ac.uk/objects/uuid:d8bbffae-8a4e-478e-ba65-0f5a5bbd66e1

[7] LOPES, Q.: *The Global Periphery: Armando de Lacerda and the Coimbra Experimental Phonetics Laboratory (1936–1979)*. Lisbon: Caleidoscópio, 2021.

[8] HAMMARSTRÖM, G.: *Phonetics: Aspects of its history*. Unpublished notes. Quoted with permission. The bibliographic references mentioned in this transcription were added to the original by the authors.

[9] BEVER, P. V.: Chronique. *Revue belge de philologie et d'histoire*, 31 (1), pp. 206–369, 1953.

[10] ROGERS, F. M.: Review: Étude de phonétique auditive sur les parlers de l'Algarve. *Romance Philology*, 8 (4), pp. 284–299, 1955.

[11] HAMMARSTRÖM, G.: *Memories of a linguist 1940–2010*. Muenchen: Lincom Europa, 2012.

[12] Arquivo do Camões, Instituto da Cooperação e da Língua, I.P. Caixa 1319, Processo 11, Documento 1.

[13] LOPES, Q. & PEREIRA, E.: Armando de Lacerda and Experimental Phonetics in the inter-war period: Scientific innovation and circulation between Portugal, Germany and Harvard. In: M. PUCHER, J. TROUVAIN & C. LOZO (Eds.): *Proceedings of the Third International Workshop on the History of Speech Communication Research*. Dresden: TUD Press, pp. 95–104, 2019.

[14] Biblioteca do Laboratório de Fonética. Faculdade de Letras de Coimbra.

[15] Göran Hammarström Family Archive (letter dated 15th February 1957).

[16] HAMMARSTRÖM, G.: *Le chromographe et le triangle tonométrique de Lacerda*. Coimbra: Laboratório de Fonética Experimental da Faculdade de Letras da Universidade de Coimbra, 1952.

[17] Göran Hammarström Family Archive (letter from Manuel Companys to Göran Hammarström dated 25th August 1958).

[18] ROSSI, N.: *Laboratório de Fonética na Bahia*. Coimbra: Laboratório de Fonética Experimental da Faculdade de Letras da Universidade de Coimbra, 1958.

[19] LACERDA, A. & ROGERS, F. M.: *Sons dependentes da fricativa palatal áfona, em Português*. Coimbra: Laboratório de Fonética Experimental da Faculdade de Letras da Universidade de Coimbra, 1939.

Author Index